Structure-Property Relations in Nonferrous Metals

Structure-Property Relations in Nonferrous Metals

Editor
_____

Vikas Saini

Structure-Property Relations in Nonferrous Metals

Edited by **Vikas Saini**

Printed in 2017

ISBN: 978-1-68117-228-6

Library of Congress Control Number: 2015936586

Contents

Preface

The book discussed on the relationships between the atomic structure, crystal structure, and microstructure of metals and their physical behavior (e.g., strength, ductility, electrical conductivity, corrosion, etc.). In Part one we present basic principles of the atomic and crystal structure, defects, and processing of metals. In Part two we describe the properties and uses of all the metallic elements except iron. We have observed that most metals texts focus primarily on iron, the dominant commercial metal.

Editor

Nanostructured Multifilamentary Carbon-Copper Composites: Fabrication, Microstructural Characterization, and Properties

Evarice Yama Nzoma, Alain Guillet,
and Philippe Pareige

Groupe de Physique des Matériaux, Université et INSA de Rouen,
Avenue de l'Université, 76801 Saint Etienne du Rouvray, France

ABSTRACT

This work is part of research on the emerging techniques to produce bulk nanostructured composites materials by severe plastic deformation and their characterization. Based on the Levi work, we present a new method to synthesize a composite wire-containing carbon-nanosized filaments (graphite and C_{60} fullerenes) embedded in a copper matrix. The originality of this process is using powder media as fiber.

Microstructures and electrical, mechanical, and thermal properties are presented.

INTRODUCTION

Plastic deformation processing is a "top-down" approach to nanostructuring metals and alloys based on the accumulation of ultralarge plastic strain to achieve microstructure refinement [1]. Various plastic-deformation techniques have been developed to produce nanostructured materials over the last two decades [2, 3]. The more common techniques are Equal-Channel Angular Pressing (ECAP) [4], High-Pressure Torsion (HPT) [5,6], and Accumulative Roll Bonding (ARB) [7, 8]. However, several variants and combination [9] of plastic deformation techniques, including accumulative repetitive deformation [3], were also developed Accumulative Spin-Bonding (ASB) [10], Equal Channel Angular Rolling (ECAR) [11, 12], Cyclic Extrusion Compression (CEC) [13], Cyclic Closed-Die Forging (CCDF) [14, 15], Linear Flow Splitting (LFS) [16], Severe Torsion Straining (STS) [17], Torsion Extrusion Process (TEP) [18], Cylinder Covered Compression (CCC) [19], Asymmetric Rolling (AR) [20, 21], and Accumulative Deep Wire-Drawing and bonding (WD) [22–24]. In this paper, we present a modified wire-drawing process.

Wire drawing is an ancient manufacturing technique of metallic profiles, bars, rods, and wires. The best known example is cold drawing of eutectoid carbon steels (0.77 wt% C): plastic deformation close to $\boxtimes$=4 leads to a considerable refinement of the pearlitic microstructure down to the nanoscale [25–27]. Codrawing and coextrusion have also enabled the rise of a new class of nanostructured metal matrix composite materials whose industrial development is not mature but got intensive research. The fabrication process is based on the Levi's technique [28]. It consists of a series of steps involving first co-drawing of bimetallic monofilamentary macrocomposite, bundling and re-stacking of the drawn elements followed by another drawing operation. Accumulative drawing and re-stacking result in to multifilamentary materials with a priori incompatible properties, such a high yield strength and good electrical conductivity [29, 30]. The archetypes of these materials are the Cu-Nb micro- or nanocomposites, used to manufacture coils generating very high magnetic fields [31, 32]. In these systems, the mechanical strength of composite in the plastic domain exceeds the "linear rule of mixtures". Except Cu-Nb, other bimetallic couples such as Cu-Fe [28, 33, 34], Cu-Ag [35, 36], Cu-Cr [37], and Cu-Ta [38, 39] are also studied. In Cu-X couples, X-fibers are used as reinforcement phase, and conducting phase is assumed by the copper matrix. The choice of the couple Cu-X is generally dictated partly by a weak mutual solubility of Cu and X to reduce phase interactions but also by the reinforcing capacity of the X element. Under these conditions, the copper-carbon (Cu-C) pair is interesting to study. Indeed these two elements are extremely immiscible [40], and therefore the composite could combine the high electrical conductivity of copper with the interesting characteristics of carbon, namely, a low coefficient of thermal expansion (CTE) and good tribological properties [41, 42].

These alloys are used in various applications such as electrical contact materials with friction [44, 45] and as heat sink for electronic components which require a material with a low CTE. Their use is also proposed to replace silver in electronic circuits [46] and as materials to confine plasma in nuclear fusion reactors [47, 48].

However, manufacturing these composites is difficult because the mutual solubility of carbon and copper [49] and the wettability of carbon by liquid copper [48, 49] are extremely low. The main manufacturing processes are the infiltration of a network of carbon fibers or graphite with copper, the deposition of copper on carbon fibers by hollow cathode and powder metallurgy. Furthermore, to our best knowledge, no process is available to manufacture nanostructured copper-carbon multifilamentary wire.

Therefore, the objective of this research is to obtain copper matrix composite wires containing millions of nanoscaled carbon filaments. We have already demonstrated the ability to make such multifilamentary composite using the Levi's technique [50]. The originality of this process [50, 51] is that carbon powder is used as a fiber unlike most of multifilament composites which are made from two bulk metallic materials. The microstructures obtained were studied at different scales and we present and discuss the resulting electrical, mechanical, and thermal properties.

MATERIALS AND EXPERIMENTAL TECHNIQUES

Materials

Copper

Copper tubes used in processing route are Cu-b2 engineering materials with 99.95 wt% purity (deoxidized low phosphorous grade). We employed tubes of different diameters: tubes A ($\emptyset_{in}.=4$ mm/$\emptyset_{out}.=8$ mm) and tubes B ($\emptyset_{in}.=10$ mm/$\emptyset_{out}.=12$ mm). In order to recrystallize the microstructure, they were heat-treated at 500°C during 3 hours under medium vacuum, followed by pickling with diluted nitric acid solution to remove oxide layers.

Carbon

Two allotropes of carbon powders were used: natural graphite and C_{60} fullerenes. Graphite consists of agglomerated particles whose average size is 2 μm [43, 50]. C_{60} fullerenes powder has a purity of 98 wt%. SEM observation of C_{60} powder reveals agglomerates of equiaxed grains with a size ranging from tens to hundreds micrometers and faceted rods structure whose width is between 1 and 10 μm and length is about 10 μm [43, 52,53]. We assume that the equiaxed grains of C_{60} result from the fragmentation of rods [54]. Before any use, and to remove organic impurities from powder purification process, 3-hour heat treatment at 150°C under medium vacuum is applied [55]. X-ray analysis reveals that it does not change the structure of the truncated icosahedron C_{60} molecules and FCC solid structure [56–59].

Fabrication Technique

The first step (i.e., stage 0, monofilamentary composite) was synthesized by filling tube A with graphite or C_{60} fullerenes powders. The powder is manually compacted with a copper rod so it is uniformly distributed in the tube. The carbon-filled tube is then drawn up to 0.7 mm diameter. It is thereupon cut into pieces of same length which are bundled and introduced into tube B. The assembly is drawn again up to 0.7 mm diameter (stage 1) as shown schematically in Figure 1. A recrystallization heat treatment (500°C, 3 h, under medium vacuum) is applied at the end of each stage or when fractures occur during the drawing operations.

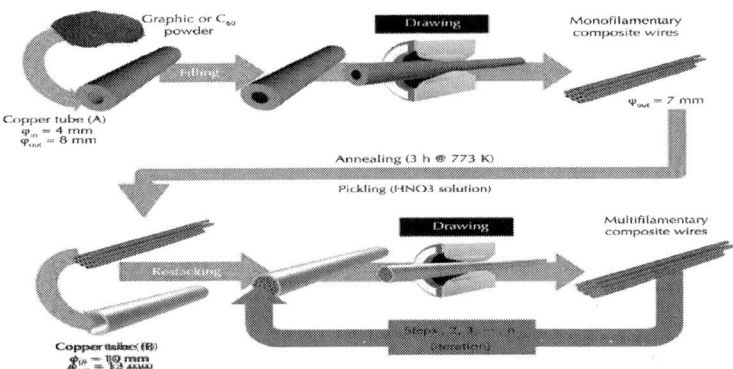

Figure 1: Schematic of Cu-C composites processing steps [43].

The evolution of graphite and C_{60} fullerenes fibers number embedded in the copper matrix according to re-stacking stages is given in Table 1. At the end of the processing Cu-C_{Gr} (final diameter: 0.8 mm with about 3.1×10^7 filaments of graphite and Cu-C_{60} wires (final diameter: 3 mm with about 1.33×10^6 filaments of C_{60}) are obtained.

Table 1: Change in Cu-C_{Gr} and Cu-C_{60} fibers number according to stage level

Stage (⊠)	Cu-C_{Gr}	Cu-C_{60}
0	1 filament	1 filament
1	110 filaments	110 filaments
2	8 800 filaments (1 1 0 × 8 0)	12 100 filaments (1 1 0 × 1 1 0)
3	704 000 filaments (8 8 0 0 × 8 0)	1.33 million of filaments (1 1 0 × 1 2 1 0 0)
4	31.7 million of filaments (7 0 4 0 0 0 × 4 5)	—

Microstructural Observations and Mechanical, Electrical, and Thermal Characterizations Techniques

The samples microstructures were characterized by optical microscopy (OM) and scanning electron microscopy (SEM) following a metallographic preparation: mounting in a bakelite resin, mechanical polishing ending with 3 μm diamond paste and chemical-mechanical polishing with a colloidal suspension of silica to which few drops of hydrogen peroxide were added. The microstructure of the copper matrix is revealed by etching few seconds with an iron chloride alcoholic solution. The OM observations were made with an Olympus microscope coupled to a CCD camera to acquire digital images enabling image analysis (AnalySIS software from the Olympus Soft Imaging System Company). The SEM images were obtained from a LEO FE1530 equipped with a field emission gun.

The deformation and recrystallized texture in metals with FCC lattices have been studied for a long time [60,61]. The main method to study crystallographic orientation is X-ray diffraction texture analysis. The texture measurements were performed using X-ray four-circle diffractometer. The samples have coin-like shape between 4 mm to 5 mm in diameter and a thickness of less than 6 mm. One side is polished to equalize the height of the carbon and copper. Then the samples were etched to remove the hardening owed by polishing operations. Due to small carbon filaments cross-section, we have obtained only copper texture measurements. The signal from the carbon was too weak to produce satisfactory pattern. The acquisition has required from 10 to 20 hours counting time.

The mechanical tests were carried out at the initial strain rates of $\dot{\varepsilon}$=3.3$\times$10^{-4}s^{-1} (cross-head speed of 2 mm/mm) at room temperature with an Instron 4483 tensile testing machine with a 2 kN load cell. Samples for tensile have been annealed 3 hours at 500°C, to obtain the same thermomechanical state to compare each stage. To measure electrical resistance, we used the four probes method using an OM21 microhmmeter from AOIP Company on annealed samples (3 h @ 500°C), to improve electrical conductivity. The resolution is 0.1 μΩ and the accuracy is about 0.03%. The coefficient of linear thermal expansion of several composites was measured between 25 and 800°C with a differential dilatometer NETZSCH DIL 402C. The experimental conditions are as follows:

- heating rate: 300°C/h;
- atmosphere: argon at 0.3 bar;
- specimen dimensions: ⌀=4 mm, ⌀=20 mm;
- number of test: 3 per measure;
- measurement uncertainty: ±5%.

MICROSTRUCTURES

Figure 2 presents Cu-C$_{Gr}$ sample micrographs from "stage 1" for different strains. The interstices between "stage 0" wires are still visible, but they disappear gradually with increasing strain. Figure 3 shows the multiscale features characteristic of these materials.

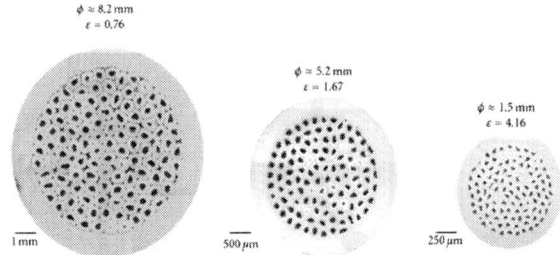

Figure 2: Optical micrographs of the Cu-C$_{Gr}$ composite cross-sections at "stage 1" at different strains [43].

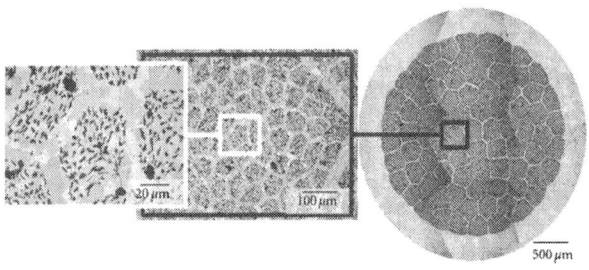

Figure 3: Optical cross-section micrographs of Cu-C$_{Gr}$ composite at "stage 3", highlighting the multiscale structure (704 000 filaments) [43].

Indeed, starting from an initial 4 mm fiber diameter and after several stages of restacking, transverse filaments dimensions became nanoscale 200 nm and 350 nm for Cu-C$_{Gr}$ and Cu-C$_{60}$, respectively. However, the morphological evolution of both Cu-C$_{Gr}$ and Cu-C$_{60}$ composites is different, especially at the highest stages. For the Cu-C$_{Gr}$ composite, three structures of graphite filaments are observed as shown in Figure 4.

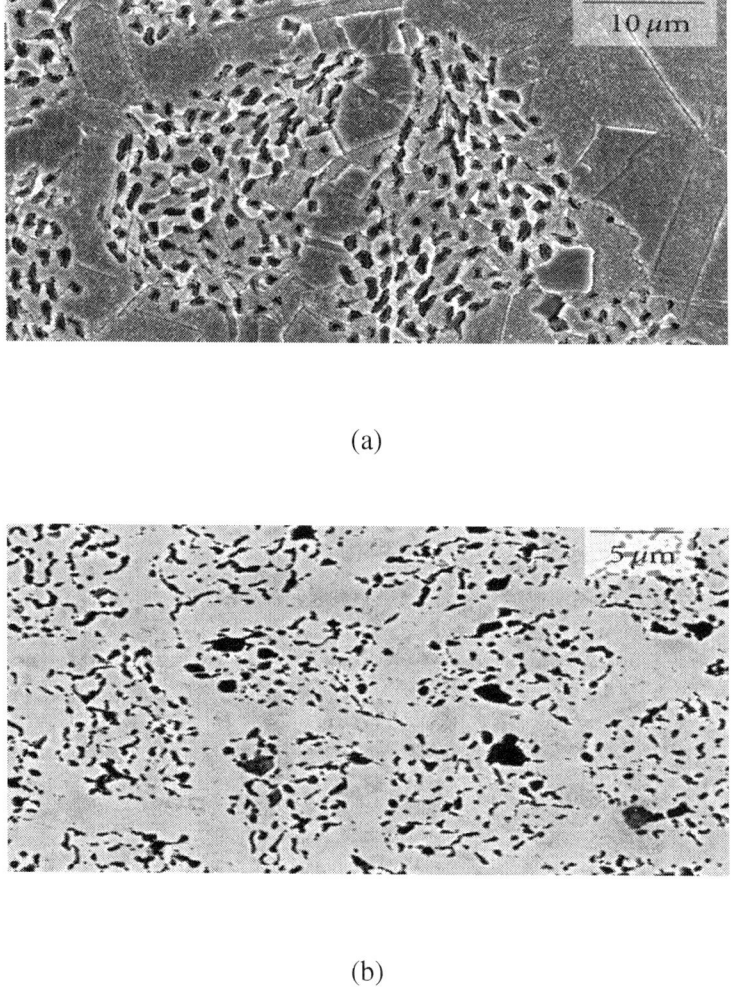

(a)

(b)

Figure 4: SEM micrographs to compare Cu-C$_{60}$ at stage 3 (a) and Cu-C$_{Gr}$ at stage 4 (b) cross-section microstructure [43].

The first graphite structure results from the agglomeration of several filaments which form large particles of several microns in diameter. The second structure consists of a network of very fine filaments of a few hundred nanometers thickness. This structure is reminiscent of the phenomenon of "curling" observed during the codeformation of bimetallic composite with different crystal structure and Young's modulus [62] (Figure4(b)). The third category is represented by filaments whose cross-section is almost circular with a diameter of several tens of nanometers. For $Cu-C_{60}$ composites, filaments are individually identifiable; however, unlike $Cu-C_{Gr}$, no agglomeration occurs, though that C_{60} filaments do not have a circular cross-section but rather an elliptical or curvilinear.

The filaments diameter is also submicrometric. The difference between both composites morphology is directly connected to the crystalline structure and therefore mechanical behavior of each carbon phase. Indeed, C_{60} is a molecular solid (truncated icosahedron molecules) crystallizing in FCC structure and the intermolecular bonds are only Van der Waals type, hence sliding is easy. The graphite structure is hexagonal with bonds between the carbon atoms of the hexagonal graphene planes and ϖ bonds between planes. Although sliding of graphene sheets with respect to another is possible, this cannot occur through the plane because the bonds are very strong. The molecular solid C_{60} therefore allows a priori an easier accommodation of mechanical stress than graphite.

In addition, etching of the composite $Cu-C_{60}$ with an iron chloride solution highlights C_{60} filaments. Indeed, after dissolution of the copper matrix, we can observe naked filaments in "bundle-like" form (Figure 5(a)). These filaments are individually identifiable and keep their longitudinal appearance (Figure 5(b)).

(a)

(b)

Figure 5: SEM images of etched Cu-C$_{60}$ composite. The submicrometer thickness and continuity of C$_{60}$ filaments are underlined. Although they are in "bundle-like" form, they are, however, individually identifiable.

This was also attempted on the Cu-C$_{Gr}$ composite but without success; during the dissolution of the copper matrix, and contrary to those of fullerenes, graphite does not retain a filamentary structure; it disperses into the etching solution. This reveals a loss of mechanical strength compared to C$_{60}$ fullerenes filaments.

Composite materials were heated to a temperature where recrystallization can occur (3 h @ 500°C), thereby strongly decreasing the number of defects caused by plastic deformation. However, as can be noted in Figure 6, the grain size of copper is not homogeneous.

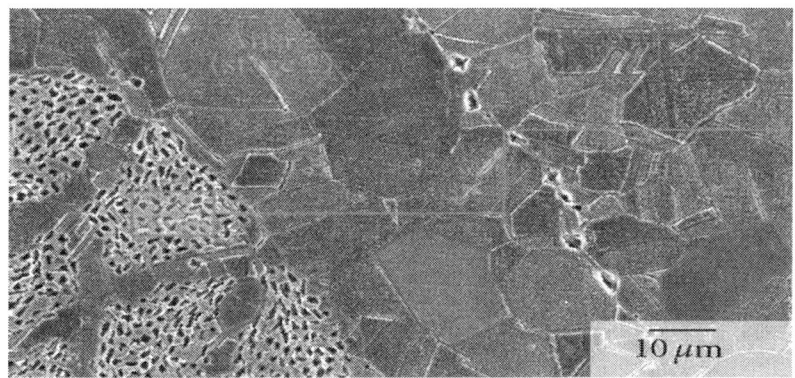

(a)

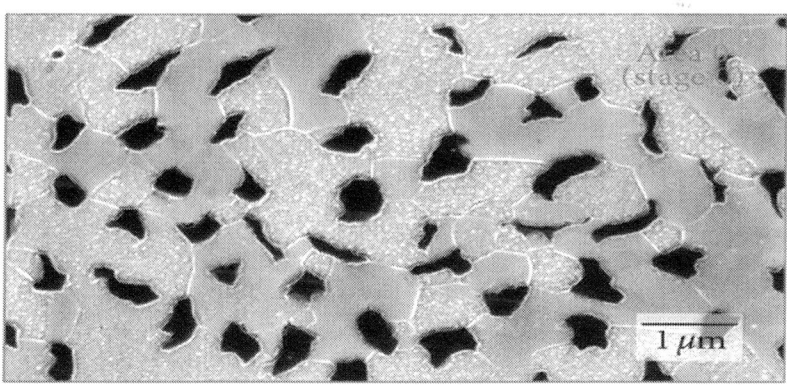

(b)

Figure 6: SEM micrographs highlighting copper matrix grains (Cu-C$_{60}$, stage 3 ⌀=5.2 mm) [43].

Indeed, depending on the processing, it varies very significantly as shown in Table 2. Grain growth is inhibited by the carbon filaments and the interfaces between the copper tubes [43].

Table 2: Mean grain size of copper matrix depending on the location of grains (Cu-C$_{60}$ and Cu-C$_{Gr}$. composites, stage 3, ⊠=5.2 mm)

Location of grain	Average grain size of copper (⊠ m)	
	Cu-C$_{60}$ stage 3	Cu-C$_{Gr}$ stage 3
Area 3	1 9 ± 6	1 9 ± 6
Area 2	1 9 ± 5	1 9 ± 5
Area 1	8 . 4 ± 2	8 . 4 ± 2
Area 0	2 . 0 ± 0 . 5	5 . 0 ± 1 . 8

Microstructure study was supplemented by X-ray diffraction analysis to extract pole figures presented in Figure 6. The literature shows that wire-drawn FCC metals such as copper have a double-directions texture ⊠111⊠ and ⊠200⊠ [63–66] along drawing axis. As it be seen in Figures 7(a) and 7(b), the ⊠111⊠ texture component dominates in cold drawn and annealed matrix of Cu-C$_{60}$ and Cu-C$_{Gr}$ composites. However, the intensity varies from one composite to another and from one stage to another. The highest degree of texturing is observed in steps 1 and 3 for Cu-C$_{Gr}$ composite materials and at step 1 for Cu-C$_{60}$ composite materials. The ⊠100⊠ component comes in lower intensity except for the composite Cu-C$_{Gr}$ in step 1.

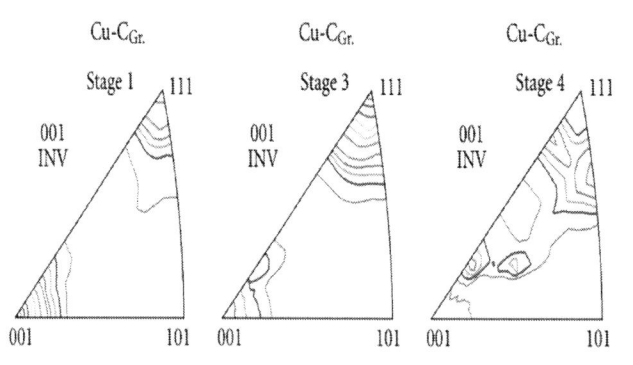 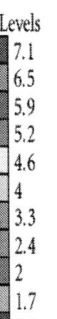

(a)

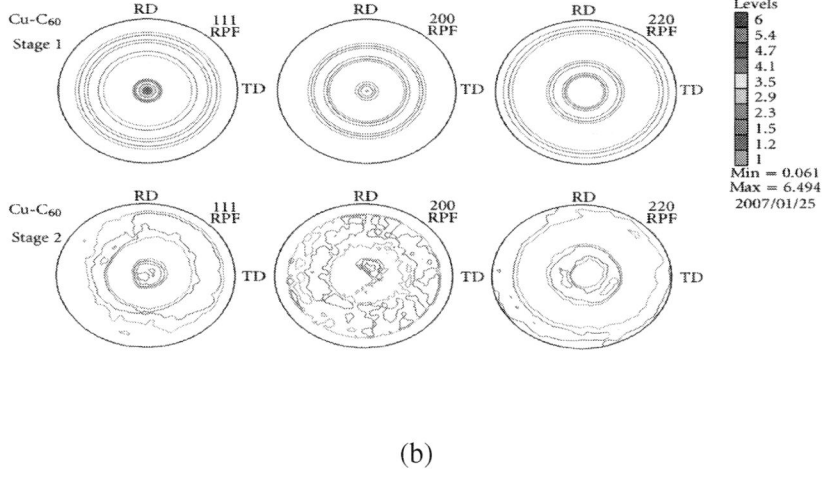

(b)

Figure 7: X-ray pole figures of Cu matrix after deformation and recrystallization showing ⬛111⬛ and ⬛100⬛ texture components measured from cross-section of rod sample. (a) Pole figures of samples Cu-C$_{Gr}$ in stages 1, 3, and 4. (b) Cu-C$_{60}$ in stages 1 and 2 pole figures. All samples were annealed 3 hours at 500°C [43].

According to several studies [67–69] intermediate recrystallization annealing and severe deformation (⬛<9) may affect and change the crystallographic orientations of deformed copper matrix. On the one hand, they showed that copper, which has undergone a deformation of ⬛<9, develops a ⬛111⬛ structural component in majority and more stable [70, 71]. On the other hand, intermediate recrystallization annealing and the application of very high deformations causing dynamic recrystallization of the Cu matrix [72] will promote the development of ⬛100⬛ component. The presence of these two types of texture in our samples shows that there has been deformation and then recrystallization of the copper matrix, although textures associated with intermediate annealed composites are low except for the composite Cu-C$_{Gr}$ stage 1. We see globally on both types of composite that an increase of the number of stages (Cu-C$_{60}$ step 2 and Cu-C$_{Gr}$ stage 4) causes a decrease of the intensity of the ⬛111⬛ texture component. Indeed, Dubois et al. [73] show that this regime is specific to nanocomposites metal and corresponds to a frustration of the normal recovery and recrystallization processes, impeded by the increased thermal resistance of Cu nanochannels.

RESULTS AND DISCUSSION

Electrical Resistivity

Cu-C_{Gr} and Cu-C_{60} composites electrical resistivity is around $2\,\mu\Omega\cdot cm$, close to Cu-b2 copper used for processing (Table 3). The presence of graphite or C_{60} fullerenes, the first being a poor conductor compared to copper, the second being an insulator [43], weakly influences the electrical conductivity of the composite. Electrical resistivities of both composite materials, Cu-C_{Gr} stage 4 (89% IACS) and Cu-C_{60} stage 3 (85% IACS), are lower than most metals and copper alloys. Only silver (106% IACS) and pure copper (101% IACS) have the lowest electrical conductivity.

Table 3: Electrical resistivities (conductivities) of Cu-C_{60} and Cu-C_{Gr} composites and Cu after recristallization heat treatment (500°C, 3 h)

Cu	Cu-C_{Gr} (stage 4)	Cu-C_{60} (stage 3)	
Surface fraction ⊠S	—	0.061	0 . 0 9 6 ± 0 . 0 0 6
Nb. filaments (×10⁶)	—	31.7	1.33
⊠ (Ω⊠c m) (%IACS)	1 . 8 7 ± 0 . 0 6 (9 2 . 2 ± 3 . 0)	1 . 9 4 ± 0 . 1 5 (8 8 . 9 ± 6 . 9)	2 . 0 3 ± 0 . 0 5 (8 4 . 9 ± 2 . 1

Mechanical Properties

The stress-strain curves of Cu-C_{60} and Cu-C_{Gr} are given in Figures 8(a) and 8(b), respectively. The tests were performed on materials annealed at 500°C during 3 hours. We deliberately choose to compare the materials properties after a recrystallization annealing, because the resistivity decreases, and because in the deformed state the material is often fragile. Furthermore, comparing the mechanical properties of the as-fabricated composite would require a precise quantification of the plastic deformation for each stage, however, the deformation during processing is inhomogeneous. Indeed, following re-stacking, there is first plastic deformation of the outer tube and filling of gaps between the wires (Figure 2), and plastic deformation of these wires occurs. Therefore, since before re-stacking, wires are annealed, the outer tube containing them undergoes greater plastic deformation.

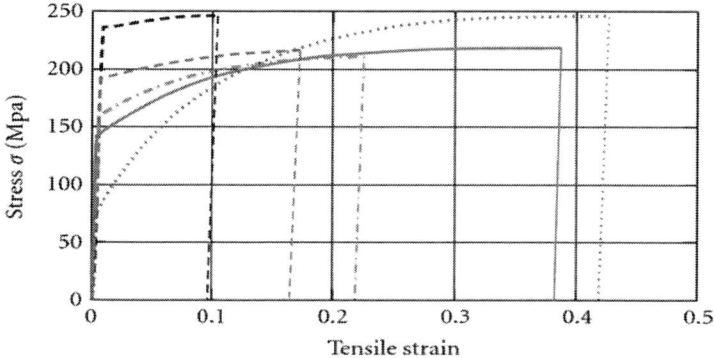

Figure 8: Engineering stress-strain curve of (a) Cu-C_{60} and (b) Cu-C_{Gr} composites from stages 0 to 3 and comparison with copper [43].

The stress-strain curves for the different re-stacking stages are shown in Figure 8. Compared to pure copper, the yield strength of Cu-C_{60} composites increases significantly while the elongation decreases.

Meanwhile the stress/strain curves for Cu-C_{Gr} composites are always below the copper one in the plastic regime. We have no certain explanation for this behaviour. In the present state of our knowledge, the comparison with the hardening of "pure" copper is difficult because even if the process of forming and heat treatments are the same, the resulting microstructures are very different. The recrystallized microstructure of copper is obviously homogeneous in terms of grain size (30 to 40 microns) [43]. That of Cu-C_{Gr} is biphasic and the grain size is heterogeneous as shown in Table 2: it varies between 2 and 19 microns, and regardless of the presence of graphite filaments, we ignore the work hardening behavior of such microstructure during tensile testing. Furthermore, the presence of an irregular and heterogeneous distribution of graphite filaments with increasing re-stacking does not allow yet a prediction of the mechanical behavior of the composites. This may explain why the stress/strain curves for Cu-C_{Gr} composites at stage 1, 2, and 3 are below the one at stage 0.

The mechanical properties determined from these stress-strain curves are shown in Table 4 but only for stage 3.

Table 4: Mechanical properties of Cu-C$_{60}$ and Cu-C$_{Gr}$ composites (stage 3) and comparison with copper

Copper	Cu-C$_{Gr}$ stage 3	Cu-C$_{60}$ stage 3	
Surface fraction ⊠	—	0.095	0.096 ± 0.006
Nb. of filaments	—	704 000	1.33 million
⊠ 0.2 % (MPa)	81 ± 11	110 ± 3	238 ± 36
⊠ (%)	41 ± 3	10 ± 4	10 ± 5
⊠ (MPa)	248 ± 11	166 ± 3	248 ± 26

The presence of carbon filaments (graphite or C$_{60}$) increases significantly the yield stress compared to pure copper, but decreases elongation at fracture. However, Cu-C$_{60}$ has the most remarkable increase since yield stress reached (238 ± 36) MPa compared to (81 ± 11) MPa for copper. The variation of yield stress depends on the mean grain size (d); the strengthening results from grain refinement according to Hall-Petch relation [74–76]:

$$\sigma(\varepsilon) = \sigma_0 + k \cdot d^{-1/2},$$

(1)

where ⊠, which falls between 0.104 and 0.25 MPa/mm$^{1/2}$ [72, 77–79], is the so-called Hall-Petch "constant," despite the fact that it is strain, temperature, and strain rate dependent. ⊠0 is the intrinsic stress (36 MPa [79]). The Hall-Petch model is based on the hypothesis that grain boundaries act as obstacles in dislocation movements. That is to say, as the grain size is reduced, dislocation movements are impeded by grain boundaries, so the yield stress of specimen increases. We have made two assumptions. First, we consider that only copper matrix contributes to the strength because carbon (graphite, C$_{60}$) filaments have a small mechanical strength compared to copper. Second, we introduced a modified Hall-Petch relation by taking into account the surface fraction of each area (see Figure 6) which has different mean grain size. Then the Hall-Petch equation becomes

$$\sigma = \sigma_0 + k \cdot \sum_{i=1}^{n} \frac{f_s^i}{\sqrt{d_i}},$$

(2)

where f_s^i is the fraction surface of grains with ⊠ size.

After calculation, we obtain a yield in the range of 63–100 MPa according on the values of ⊠. However, this result, comparable to that of the $Cu-C_{Gr}$ composite at stage 3, is below that experimental value which is 110±3 MPa. In the case of $Cu-C_{60}$ composite, the values of the theoretical yield strengths range from 67 MPa to 112 MPa (with an average of 86±23 MPa). The experimental value is nearly three times greater than calculated one.

The application of this modified Hall-Petch relationship shows that grain refinement does not fully explain this increase [43].

Other contributions should be considered especially the influence of matrix texture. It was demonstrated that even a moderate texture can result in significant anisotropy in the plastic behavior of Cu matrix [80]. Also a beginning of polymerization of filaments C_{60} [81, 82] and structure change can slightly contribute to mechanical reinforcement. A contribution from the Cu-C interface is unlikely given the low solubility of carbon and the absence of copper-carbon compounds [40].

Thermal Properties

The average coefficient of thermal expansion between 25 and 100°C (CTE) of $Cu-C_{60}$ and $Cu-C_{Gr}$ composite is given in Table 5. With the exception of the composite $Cu-C_{Gr}$ stage 2, the presence of graphite or fullerenes decreases significantly CTE compared to the pure copper (CTE=17×10^{-6} °C^{-1}) [83].

Table 5: Linear coefficient of thermal expansion of Cu-C$_{Gr}$ and Cu-C$_{60}$ composites in the temperature range (25°C–100°C) depending on the number of filaments and carbon fraction surface

		Stage 1	Stage 2	Stage 3	Stage 4
Cu-C$_{Gr.}$	Nb of filaments of C$_{Gr}$	110	8800	704000	31.7 million
	Surface fraction of C$_{Gr}$	0.092	0.074	0.095	0.061
	CTE$_{mes}$ ($\times10^{-6°}$ C−1)	13.7	16.8	15.1	11
Cu-C$_{60}$	Nb of filaments of C$_{60}$	—	12100	1.33 millions	—
	Surface fraction of C$_{60}$	—	0.11 ± 0.05	0.096 ± 0.005	—
	CDT$_{mes}$ ($\times10^{-6°}$ C−1)	—	12.4	13.3	—

The decrease of CTE is the largest for Cu-C$_{Gr}$ at stage 4, (11×10^{-6} °C^{-1}). It is also surprising that the CTE of Cu-C$_{60}$ is well below the CTE of the components, namely, 17×10^{-6} °C^{-1} for copper and 20×10^{-6} °C^{-1} for C$_{60}$. Therefore, we may assume that successive annealing and cumulate deformation can lead to a polymerization of C$_{60}$ with a change of structure and properties [81, 82].

CONCLUSIONS AND PERSPECTIVES

In this work we demonstrate that the manufacturing method of bimetallic composite multifilamentary wire by codrawing and re-stacking is applicable to the development of Copper-Carbon micro- or nanocomposite wires where the matrix is the metallic phase and the second phase is a carbon powder. The fabrication process combines simultaneously plastic deformation and powder compaction. The successive wire-drawing and re-stacking lead to multi-scale microstructure of the metal matrix. The diameter of carbon filaments is greatly reduced. Thus, the Cu-C$_{Gr}$ wire has more than 3.1×10^7 graphite filaments with a diameter of 200 nm, while that of Cu-C$_{60}$ contains more than a million of filaments with 350 nm in diameter.

This process enables development of materials with properties better or close to copper. Indeed, the yield stress is significantly increased compared with that of copper, while the elongation decreases dramatically. The electrical conductivity is weakly influenced by the presence of carbon (graphite or fullerenes), but the coefficient of linear thermal expansion decreases notably.

Cumulative co-drawing and bundling can be extended to other metal-carbon systems like aluminum-carbon nanotubes and nonmetallic, such as metal-polymer [84]. We also demonstrated that after drawing and etching, this process enables to obtain naked C$_{60}$ fullerenes filaments with nano size section. These filaments, thus highlighted, can be used as fibers which could be coated by metal, polymeric, or ceramic matrix.

ACKNOWLEDGMENTS

This work is supported by the "Groupe de Physique des Matériaux" laboratory in Nanostructures and Phase Transformation Research Team. The authors are grateful to CRITT Metal 2T of Nancy for linear thermal expansion analysis and "Materials and Processes Sciences Laboratory" (LSMP) in Paris XIII University for the texture measurements.

REFERENCES

1. A. Misra and L. Thilly, "Structural metals at extremes," MRS Bulletin, vol. 35, no. 12, pp. 965–972, 2010.

2. Y. Zhu, R. Z. Valiev, T. G. Langdon, N. Tsuji, and K. Lu, "Processing of nanostructured metals and alloys via plastic deformation," MRS Bulletin, vol. 35, no. 12, pp. 977–981, 2010.

3. A. Azushima, R. Kopp, A. Korhonen et al., "Severe plastic deformation (SPD) processes for metals,"CIRP Annals. Manufacturing Technology, vol. 57, no. 2, pp. 716–735, 2008.

4.   R. Z. Valiev and T. G. Langdon, "Principles of equal-channel angular pressing as a processing tool for grain refinement," Progress in Materials Science, vol. 51, no. 7, pp. 881–981, 2006.

5.   P. W. Bridgman, "Effects of high shearing stress combined with high hydrostatic pressure," Physical Review, vol. 48, no. 10, pp. 825–847, 1935.

6.   S. Erbel, "Mechanical properties and structure of extremely strain-hardened copper," Metals Technology, vol. 6, no. 12, pp. 482–486, 1979.

7.   N. Tsuji, Y. Saito, S. H. Lee, and Y. Minamino, "ARB (accumulative roll-bonding) and other new techniques to produce bulk ultrafine grained materials," Advanced Engineering Materials, vol. 5, no. 5, pp. 338–344, 2003.

8.   N. Tsuji, "Nature of deformation behavior in ultrafine grained metallic materials," in Proceedings of the 3rd International Conference on Nanomaterials by Severe Plastic Deformation (Nano SPD3'05), Fukuoka, Japan, 2005.

9.   D.-F. Zhang, H.-J. HU, F. S. Pan, M.-B. Yang, and J.-P. Zhang, "Numerical and physical simulation of new SPD method combining extrusion and equal channel angular pressing for AZ31 magnesium alloy,"Transactions of Nonferrous Metals Society of China, vol. 20, no. 3, pp. 478–483, 2010.

10.   M. S. Mohebbi and A. Akbarzadeh, "Accumulative spin-bonding (ASB) as a novel SPD process for fabrication of nanostructured tubes," Materials Science and Engineering A, vol. 528, no. 1, pp. 180–188, 2010.

11.   J. C. Lee, H. K. Seok, J. Y. Suh, J. H. Han, and Y. H. Chung, "Structural evolution of a strip-cast al alloy sheet processed by continuous equal-channel angular pressing," Metallurgical and Materials Transactions A, vol. 33, no. 3, pp. 665–673, 2002.

12.   J. C. Lee, J. Y. Suh, and J. P. Ahn, "Work-softening behavior of the ultrafine-grained Al alloy processed by high-strain-rate, dissimilar-channel angular pressing," Metallurgical and Materials Transactions A, vol. 34, no. 3, pp. 625–632, 2003.

13.   A. Korbel, M. Richert, and J. Richert, "The effects of very high cumulative deformation onstructure and mechanical properties of aluminium," in Proceedings of the 2nd RISO International

Symposium on Metallurgy and Material Science Roskilde, pp. 445–450, Roskilde, Denmark, 1981.

14. A. K. Ghosh and W. Huang, Investigations and Applications of Severe Plastic Deformation, vol. 80 ofNATO Science Series, T. C. Lowe and R. Z. Valiev, Eds., Kluwer Academic Publications, Boston, Mass, USA, 2000.

15. R. Kuziak, W. Zalecki, M. Pietrzyk, and S. Wçglarczyk, "New possibilities of achieving ultrafine grained microstructure in metals and alloys employing MaxStrain technology," Bulk and Graded Nanometals, vol. 101-102, pp. 43–48, 2005.

16. P. Groche, D. Vucic, and M. Jöckel, "Basics of linear flow splitting," Journal of Materials Processing Technology, vol. 183, no. 2-3, pp. 249–255, 2007.

17. K. Nakamura, K. Neishi, K. Kaneko, M. Nakagaki, and Z. Horita, "Development of severe torsion straining process for rapid continuous grain refinement," Materials Transactions, vol. 45, no. 12, pp. 3338–3342, 2004.

18. S. Mizunuma, "Large straining behavior and microstructure refinement of several metals by torsion extrusion process," Materials Science Forum, vol. 503-504, pp. 185–190, 2006.

19. X. Zhao, T. F. Jing, Y. W. Gao, J. F. Zhou, and W. Wang, "A new SPD process for spheroidal cast iron,"Materials Letters, vol. 58, no. 19, pp. 2335–2339, 2004.

20. J. H. Jiang, Y. Ding, F. Q. Zuo, and A. D. Shan, "Mechanical properties and microstructures of ultrafine-grained pure aluminum by asymmetric rolling," Scripta Materialia, vol. 60, pp. 905–908, 2009.

21. Y. H. Ji and J. J. Park, "Development of severe plastic deformation by various asymmetric rolling processes," Materials Science and Engineering A, vol. 499, no. 1-2, pp. 14–17, 2009.

22. K. Hono, M. Ohnuma, M. Murayama, S. Nishida, A. Yoshie, and T. Takahashi, "Cementite decomposition in heavily drawn pearlite steel wire," Scripta Materialia, vol. 44, no. 6, pp. 977–983, 2001.

23. K. Hanazaki, N. Shigeiri, and N. Tsuji, "Change in microstructures and mechanical properties during deep wire drawing of copper," Materials Science and Engineering A, vol. 527, no. 21–22, pp. 5699–5707, 2010.

24. N. Tsuji and K. Hanazaki, "Nanostructure formation during deep wire-drawing of copper," Materials Science Forum, vol. 654–656, pp. 1201–1204, 2010.

25. P. Watté, J. Van Humbeeck, E. Aernoudt, and I. Lefever, "Strain ageing in heavily drawn eutectoid steel wires," Scripta Materialia, vol. 34, no. 1, pp. 89–95, 1996.

26. A. V. Korznikov, Y. V. Ivanisenko, D. V. Laptionok, I. M. Safarov, V. P. Pilyugin, and R. Z. Valiev, "Influence of severe plastic deformation on structure and phase composition of carbon steel,"Nanostructured Materials, vol. 4, no. 2, pp. 159–167, 1994.

27. J. Languillaume, G. Kapelski, and B. Baudelet, "Cementite dissolution in heavily cold drawn pearlitic steel wires," Acta Materialia, vol. 45, no. 3, pp. 1201–1212, 1997.

28. F. P. Levi, "Permanent magnets obtained by drawing compacts of parallel iron wires," Journal of Applied Physics, vol. 31, no. 8, pp. 1469–1471, 1966.

29. A. M. Russell, L. S. Chumbley, andY.Tian, "Deformation processed metal-metal composites," Advanced Engineering Materials, vol. 2, no. 1-2, pp. 11–22, 2000.

30. D. Raabe, F. Heringhaus, U. Hangen, and G. Gottstein, "Investigation of a Cu-20 mass% Nb in situ composite part I: fabrication, microstructure and mechanical properties," Zeitschrift fuer Metallkunde, vol. 86, no. 6, pp. 405–415, 1995.

31. L. Thilly, M. Véron, O. Ludwig, F. Lecouturier, J. P. Peyrade, and S. Askénazy, "High-strength materials: in-situ investigations of dislocation behaviour in Cu-Nb multifilamentary nanostructured composites,"Philosophical Magazine A, vol. 82, no. 5, pp. 925–942, 2002.

32. L. Thilly, M. Véron, O. Ludwig, and F. Lecouturier, "Deformation mechanism in high strength Cu/Nb nanocomposites," Materials Science and Engineering A, vol. 309-310, pp. 510–513, 2001.

33. X. Sauvage, F. Wetscher, and P. Pareige, "Mechanical alloying of Cu and Fe induced by severe plastic deformation of a Cu-Fe composite," Acta Materialia, vol. 53, no. 7, pp. 2127–2135, 2005.

34. Y. S. Go and W. A. Spitzig, "Strengthening in deformation-processed Cu-20% Fe composites," Journal of Materials Science, vol. 26, no. 1, pp. 163–171, 1991.

35. D. W. Yao and L. Meng, "Effects of solute, temperature and strain on the electrical resistivity of Cu-Ag filamentary composites," Physica B, vol. 403, no. 19-20, pp. 3384–3388, 2008.

36. Y. Sakai and H. J. Schneider-Muntau, "Ultra-high strength, high conductivity Cu-Ag alloy wires," Acta Materialia, vol. 45, no. 3, pp. 1017–1023, 1997.

37. C. Masuda and Y. Tanaka, "Fatigue properties of Cu-Cr in situ composite," International Journal of Fatigue, vol. 28, no. 10, pp. 1426–1434, 2006.

38. K. Sakurai, Y. Yamada, C. H. Lee, T. Fukunaga, and U. Mizutani, "Solid state amorphization in the CuTa alloy system," Materials Science and Engineering A, vol. 134, no. C, pp. 1414–1417, 1991.

39. L. Thilly, J. Colin, F. Lecouturier, J. P. Peyrade, J. Grilhé, and S. Askénazy, "Interface instability in the drawing process of copper/tantalum conductors," Acta Materialia, vol. 47, no. 3, pp. 853–857, 1999.

40. T. B. Massalski and H. Okamoto, Binary Alloy Phase Diagrams, ASM International, Ohio, USA, 2nd edition, 1990.

41. S. F. Moustafa, S. A. El-Badry, A. M. Sanad, and B. Kieback, "Friction and wear of copper-graphite composites made with Cu-coated and uncoated graphite powders," Wear, vol. 253, no. 7-8, pp. 699–710, 2002. · ·

42. T. Oku, A. Kurumada, T. Sogabe, T. Oku, T. Hiraoka, and K. Kuroda, "Effects of titanium impregnation on the thermal conductivity of carbon/copper composite materials," Journal of Nuclear Materials, vol. 257, no. 1, pp. 59–66, 1998.

43. E. Y. Nzoma, Elaboration par déformation plastique intense et caractérisation de composites multifilamentaires nanostructurés cuivre-carbone, Ph.D. thesis, University of Rouen, France, 2009.

44. J. M. García-Márquez, N. Antón, A. Jimenez, M. Madrid, M. A. Martinez, and J. A. Bas, "Viability study and mechanical characterisation of copper-graphite electrical contacts produced by adhesive joining,"Journal of Materials Processing Technology, vol. 143-144, no. 1, pp. 290–293, 2003. · ·

45.   D. H. He and R. Manory, "A novel electrical contact material with improved self-lubrication for railway current collectors," Wear, vol. 249, no. 7, pp. 626–636, 2001. · View at Google Scholar·

46.   S. Dorfman and D. Fuks, "Diffusivity of carbon in the copper matrix. Influence of alloying," Composites Part A, vol. 27, no. 9, pp. 697–701, 1996. · ·

47.   C. H. Stoessel, J. C. Withers, C. Pan, D. Wallace, and R. O. Loutfy, "Improved hollow cathode magnetron deposition for producing high thermal conductivity graphite-copper composite," Surface and Coatings Technology, vol. 76-77, no. 2, pp. 640–644, 1995. · ·

48.   P. Appendino, M. Ferraris, V. Casalegno, M. Salvo, M. Merola, and M. Grattarola, "Direct joining of CFC to copper," Journal of Nuclear Materials. B, vol. 329-333, no. 1-3, pp. 1563–1566, 2004. · ·

49.   G. A. López and E. J. Mittemeijer, "The solubility of C in solid Cu," Scripta Materialia, vol. 51, no. 1, pp. 1–5, 2004. · ·

50.   A. Guillet, E. Y. Nzoma, and P. Pareige, "A new processing technique for copper-graphite multifilamentary nanocomposite wire: microstructures and electrical properties," Journal of Materials Processing Technology, vol. 182, no. 1-3, pp. 50–57, 2007. · ·

51.   P. Tautzenberger, L. Tillmann, H. Wilhelm, and D. Stoeckel, "Deformation and properties of fibre reinforced composite materials," Wire, vol. 28, no. 2, pp. 65–71, 1979.

52.   F. Michaud, M. Barrio, S. Toscani, V. Agafonov, H. Szwarc, and R. Céolin, "Solid-state studies on solvated [60] fullerene crystals grown from trichlorethylene," Fullerene Science and Technology, vol. 5, no. 7, pp. 1645–1650, 1997.

53.   R. Ceolin, "A fullerene or footballene (C_{60}) is a molecule made up of 60 or 70 atoms of carbon located at the vertices of hexagons and pentagons to form a closed cage that looks similar to a football," CNRS-INFO (URA1104—Physical Chemistry of Raw Materials, ORSAY), vol. 233, 1991.

54.   G. Beaucage, J. E. Mark, G. T. Burns, and H. D.-Wu, Nanostructured Powders and Their Industrial Application, Proceedings of the Materials Research Society, Materials Research Society, San Francisco Calif, USA, 1998.

55. E. E. B. Campbell , "Fullerene sources," in Fullerene Collision Reactions, Developments in Fullerene Science, 1st edition, 2004.

56. W. Kratschmer, L. D. Lamb, K. Fostiropoulos, and D. R. Huffman, "Solid C_{60}: a new form of carbon,"Nature, vol. 347, no. 6291, pp. 354–358, 1990. ··

57. J. L. Wragg, J. E. Chamberlain, H. W. White, W. Krätschmer, and D. R. Huffman, "Scanning tunnelling microscopy of solid C_{60}/C_{70}," Nature, vol. 348, no. 6302, pp. 623–624, 1990.

58. P. A. Heiney, J. E. Fischer, A. R. McGhie et al., "Orientational ordering transition in solid C60," Physical Review Letters, vol. 66, no. 22, pp. 2911–2914, 1991. ··

59. K. Prassides and S. Margadonna, "Structures of Fullerene-based solids," in Fullerenes: Chemistry, Physics, and Technology, K. M. Kadish and R. S. Ruoff, Eds., p. 555, John Wiley & Sons, 2000.

60. H. R. Wenk and U. F. Kocks, "Representation of orientation distributions," Metallurgical transactions. A, vol. 18, no. 6, pp. 1083–1092, 1987.

61. J. S. Kallend, U. F. Kocks, A. D. Rollett, and H. R. Wenk, "Operational texture analysis," Materials Science and Engineering A, vol. 132, no. C, pp. 1–11, 1991.

62. X. Quelennec, Nanostructuration d'un composite Cu-Fe par déformation intense : vers un mélange à l'échelle atomique, Ph.D. thesis, University of Rouen, France, 2008.

63. J. D. Verhoeven, W. A. Spitzig, L. L. Jones et al., "Development of deformation processed copper-refractory metal composite alloys," Journal of Materials Engineering, vol. 12, no. 2, pp. 127–139, 1990. ··

64. E. Snoeck, F. Lecouturier, L. Thilly et al., "Microstructural studies of in situ produced filamentary Cu/Nb wires," Scripta Materialia, vol. 38, no. 11, pp. 1643–1648, 1998.

65. D. Raabe and U. Hangen, "Introduction of a modified linear rule of mixtures for the modelling of the yield strength of heavily wire drawn in situ composites," Composites Science and Technology, vol. 55, no. 1, pp. 57–61, 1995.

66. U. Hangen and D. Raabe, "Modelling of the yield strength of a heavily wire drawn Cu-20%Nb composite by use of a modified linear rule of mixtures," Acta Metallurgica Et Materialia, vol. 43, no. 11, pp. 4075–4082, 1995. ··

67. E. N. Popova, V. V. Popov, E. P. Romanov, N. E. Hlebova, and A. K. Shikov, "Effect of deformation and annealing on texture parameters of composite Cu-Nb wire," Scripta Materialia, vol. 51, no. 7, pp. 727–731, 2004. ··

68. J. Chen, W. Yan, W. Li, J. Miao, and X.-H. Fan, "Texture evolution and its simulation of cold drawing copper wires produced by continuous casting," Transactions of Nonferrous Metals Society of China, vol. 21, no. 1, pp. 152–158, 2011. · View at Google Scholar

69. J. Chen, W. Yan, X. Wang, and X. Fan, "Microstructure evolution of single crystal copper wires in cold drawing," Science in China, Series E: Technological Sciences, vol. 50, no. 6, pp. 736–748, 2007. ··

70. J. B. Dubois, L. Thilly, P. O. Renault, F. Lecouturier, and M. Di Michiel, "Thermal stability of nanocomposite metals: in situ observation of anomalous residual stress relaxation during annealing under synchrotron radiation," Acta Materialia, vol. 58, no. 19, pp. 6504–6512, 2010. ··

71. A. Kauffmann, J. Freudenberger, D. Geissler et al., "Severe deformation twinning in pure copper by cryogenic wire drawing," Acta Materialia, vol. 59, no. 20, pp. 7816–7823, 2011. ·

72. S. Lefebvre, B. Devincre, and T. Hoc, "Simulation of the Hall-Petch effect in ultra-fine grained copper,"Materials Science and Engineering A, vol. 400-401, no. 1-2, pp. 150–153, 2005. ··

73. J. B. Dubois, L. Thilly, P. O. Renault, F. Lecouturier, and M. Di Michiel, "Thermal stability of nanocomposite metals: in situ observation of anomalous residual stress relaxation during annealing under synchrotron radiation," Acta Materialia, vol. 58, no. 19, pp. 6504–6512, 2010. ··

74. E. O. Hall, "The deformation and ageing of mild steel: III Discussion of results," Proceedings of the Physical Society. Section B, vol. 64, no. 9, article no. 303, pp. 747–753, 1951. ··

75. N. J. Petch, "The cleavage strength of polycrystals," Journal of the Iron and Steel Institute, vol. 74, p. 25, 1953.

76. T. H. Courtney, Mechanical Behavior of Materials, McGraw-Hill, Singapore, 2000.

77. N. Hansen, "Hall-petch relation and boundary strengthening," Scripta Materialia, vol. 51, no. 8, pp. 801–806, 2004. ··

78. N. Hansen and B. Ralph, "The strain and grain size dependence of the flow stress of copper," Acta Metallurgica, vol. 30, no. 2, pp. 411–417, 1982.

79. A. Lasalmonie and J. L. Strudel, "Influence of grain size on the mechanical behaviour of some high strength materials," Journal of Materials Science, vol. 21, no. 6, pp. 1837–1852, 1986. ··

80. S. R. Agnew and J. R. Weertman, "The influence of texture on the elastic properties of ultrafine-grain copper," Materials Science and Engineering A, vol. 242, no. 1-2, pp. 174–180, 1998.

81. Y. K. Kwon, S. Berber, and D. Tománek, "Thermal contraction of carbon fullerenes and nanotubes,"Physical Review Letters, vol. 92, no. 1, Article ID 015901, pp. 159011–159014, 2004.

82. V. I. Zubov, N. P. Tretiakov, J. N. Teixeira Rabelo, and J. F. Sanchez Ortiz, "Calculations of the thermal expansion, cohesive energy and thermodynamic stability of a Van der Waals crystal—fullerene C_{60},"Physics Letters A, vol. 194, no. 3, pp. 223–227, 1994.

83. R. Valdiviez, D. Schrage, H. Haagenstad, and J. Szalczinger, "The thermal expansion of some common Linac materials," in Proceedings of the 21st International Linear Accelerator Conference of LINAC, TH474, Gyeongju, Republic of Korea, 2002.

84. A. Guillet, E. Dargent, L. Delbreilh, P. Pareige, and J. M. Saiter, "Fabrication and characterization of multi-filament copper matrix-polyethylene fibres composite wire," Composites Science and Technology, vol. 69, no. 7-8, pp. 1218–1224, 2009. ··

Experimental Investigation of the Corrosion Behavior of Friction Stir Welded AZ61A Magnesium Alloy Welds under Salt Spray Corrosion Test and Galvanic Corrosion Test Using Response Surface Methodology

A. Dhanapal[1], S. Rajendra Boopathy[2], V. Balasubramanian[3], K. Chidambaram[1], and A. R. Thoheer Zaman[1]

[1]Department of Mechanical Engineering, Sri Ramanujar Engineering College, Vandalur, Chennai, Tamil Nadu 600 048, India

[2]Department of Mechanical Engineering, College of Engineering, Anna University, Chennai 600 025, India

[3]Center for Materials Joining & Research (CEMAJOR), Department of Manufacturing Engineering, Annamalai University, Annamalai Nagar, Chidambaram 608 002, India

ABSTRACT

Extruded Mg alloy plates of 6 mm thick of AZ61A grade were butt welded using advanced welding process and friction stir welding (FSW) processes. The specimens were exposed to salt spray conditions and immersion conditions to characterize their corrosion rates on the effect of pH value, chloride ion concentration, and corrosion time. In addition, an attempt was made to develop an empirical relationship to predict the corrosion rate of FSW welds in salt spray corrosion test and galvanic corrosion test using design of experiments. The corrosion morphology and the pit morphology were analyzed by optical microscopy, and the corrosion products were examined using scanning electron microscope and X-ray diffraction analysis. From this research work, it is found that, in both corrosion tests, the corrosion rate decreases with the increase in pH value, the decrease in chloride ion concentration, and a higher corrosion time. The results show the usage of the magnesium alloy for best environments and suitable applications from the aforementioned conditions. Also, it is found that AZ61A magnesium alloy welds possess low-corrosion rate and higher-corrosion resistance in the galvanic corrosion test than in the salt spray corrosion test.

INTRODUCTION

Magnesium alloys have received extensive recognition due to their excellent physical properties, including light weight, high strength/weight ratio, high thermal conductivity, and good electromagnetic shielding characteristics; thus, become promising candidates to replace steel and aluminum alloys in many structural and mechanical applications due to their attractive mechanical and metallurgical properties [1, 2]. The joining of magnesium components made from this alloy is, however, still limited. Unfortunately, conventional fusion welding of magnesium alloys often produces porosity and hot cracks in the welded joint. This deteriorates both the mechanical properties and corrosion resistance [3, 4]. Hence, it will be of extreme benefit if a solid state joining process, that is, one which avoids bulk melting of the base materials, hot cracking, and porosity, can be developed and carried out for the joining of magnesium alloys.

FSW is a solid state welding process without emission of ration or dangerous fumes, and it avoids the formation of solidification defects like hot cracking and porosity. Moreover, it significantly improved the weld properties and had been extensively applied in the joining of magnesium alloys [5]. The application of Mg alloy in the structural members is still limited due to its conventional fusion welding resulting in many solidifications related problems such as hot cracking, porosity, alloy segregation, and partial melting zone. To overcome the previously said problems, FSW process had been used which is a solid state autogenous process, and, hence, there are no melting and solidification defects.

However, the corrosion resistance of the Mg-based alloys is generally inadequate due to the low-standard electrochemical potential $-2.37\,V$ compared to the (SHE) standard hydrogen electrode [6], and this limits the range of applications for Mg and its alloys. Therefore, the study of corrosion behavior of magnesium alloys in active media, especially those containing aggressive ions, is crucial to the understanding the corrosion mechanisms and, hence, to improving the corrosion resistance under various service conditions. Salt spray testing is the main technique for corrosion studies, which was employed in this research in an effort to expose the AZ61 Mg alloy to an environment similar to that experienced by automotive engine blocks [7]. It is well known that Mg alloys are susceptible to corrosion such as pitting and stress cracking corrosion (SCC). Major studies show that the SCC susceptibility of Mg alloys is increased in solutions containing chloride [8, 9]. The galvanic corrosion of magnesium using a (GCA) galvanic corrosion assembly which systematically investigates the influence of cathode materials the distance between anode and cathode, also for the anode/cathode area ratio. This study identified important effects such as the "alkalization," "passivation," poisoning, and shortcuts effect as well as the effectiveness of an insulating spacer in reduced galvanic corrosion [10]. It is well known that Mg alloys are susceptible to corrosion such as pitting and stress cracking corrosion (SCC). More studies show that the SCC susceptibility of Mg alloys is increased in solutions containing chloride [11, 12]. The welding process inevitably causes changes in the original microstructure of the alloy due to welding thermal cycles. These microstructural changes can affect the localized corrosion behavior of the alloy [13]. Thus, the present study contributed towards the galvanic effect of the friction stir welded AZ61A magnesium alloy weld in contact with AZ61A Mg base metal couple. Galvanic corrosion, which is originally defined as the enhanced corrosion between two or more electrically connected metals. It is one of the most common forms of corrosion considering the real world engineering structures [14].

This research focused on comparing salt spray testing with galvanic corrosion testing, which are the two main techniques for the corrosion studies in an effort to expose the magnesium alloy and its welds to environments similar to those environments experienced for automotive and structural applications. Moreover, galvanic corrosion has never been investigated using identical couple electrodes; so in this present investigation, a new method is enhanced to predict the galvanic corrosion of FSW AZ61A Magnesium alloy. From the literature review, it was understood that most of the published information on corrosion behavior of Mg alloys was focused on general corrosion and pitting corrosion of unwelded base alloys. Very few investigations have been conducted so far on corrosion behavior of FSW joints of Mg alloys. The aim of this research is to investigate the occurrence of salt spray corrosion in FSW welds and galvanic corrosion in weld zone with parent alloy of AZ61A Mg alloy. Hence, the present investigation was carried out to study the effect of pH value, chloride ion concentration, and corrosion time on corrosion rate of AZ61A magnesium alloy welds and the galvanic couple.

EXPERIMENTAL PROCEDURE

Test Materials

The material used in this study was an AZ61A magnesium alloy in the form of an extruded condition and supplied in plates of 6 mm thickness. The chemical composition and mechanical properties of the base metal are presented in Tables 1(a) and 1(b). The optical micrograph of the base metal is shown in Figure 1.

Table 1: (a) chemical composition (wt%) of AZ61A Mg alloy and (b) mechanical properties of AZ61A Mg alloy

(a)

Al	Zn	Mn	Mg
5.45	1.26	0.17	Balance

(b)

Yield strength (MPa)	Ultimate tensile strength (MPa)	Elongation (%)	Vickers hardness at 0.05 kg load (Hv)
177	272	8.40	57

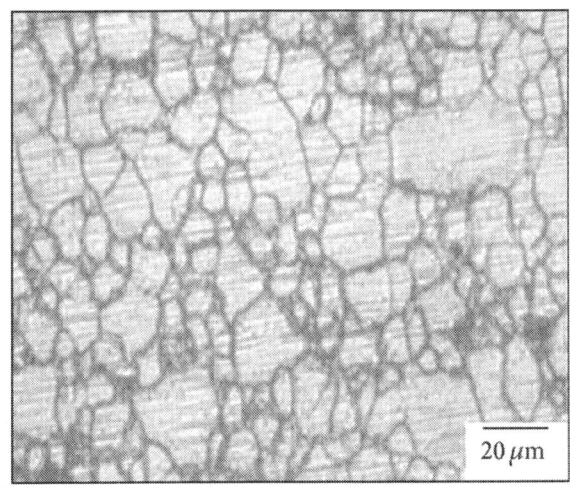

Figure 1: Optical micrograph of AZ61A base metal.

Fabricating the Joints and Preparing the Specimens

The plate was cut to a required size (300 mm × 150 mm) by power hacksaw followed by milling. A square-butt joint configuration was prepared to fabricate the joints. The initial joint configuration was obtained by securing the plates in position using mechanical clamps. The direction of welding was normal to the extruded direction. Single pass welding procedure was followed to fabricate the joints. A nonconsumable tool made of high carbon steel was used to fabricate joints. An indigenously designed and developed computer numerical controlled friction stir welding machine (22 kW; 4000 RPM; 60 kN) was used to fabricate joints. The FSW parameters were optimized by conducting trial runs, and the welding conditions which produced defect-free joints were taken as optimized welding conditions. The optimized welding conditions used to fabricate the joints in this investigation are presented in Table 2. From the base metal and welded joints, the corrosion test specimens were sliced to the dimensions of 50 mm × 16 mm × 6 mm shown in Figure 2. The specimens were ground with 500#, 800#, 1200#, 1500# grit SiC paper. Finally, it was cleaned with acetone and washed in distilled water then dried by warm flowing air. The optical micrograph of the stir zone of the FSW joint of AZ61A magnesium alloy is shown in Figure 3.

Table 2: Optimized welding conditions and process parameters used to fabricate the joints

Rotational speed (rpm)	Welding speed (mm/min)	Axial force (kN)	Tool shoulder diameter (mm)	Pin diameter (mm)	Pin length (mm)	Pin profile
1000	75	3	18	6	5	Left hand thread of 1 mm pitch

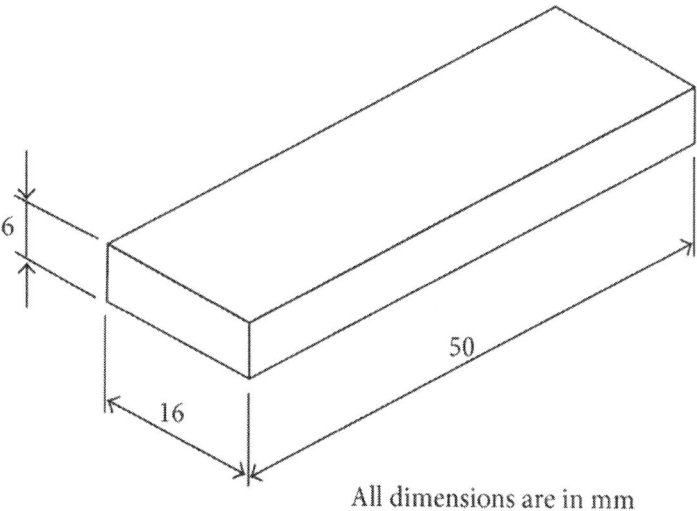

All dimensions are in mm

Figure 2: Dimensions of corrosion test specimen.

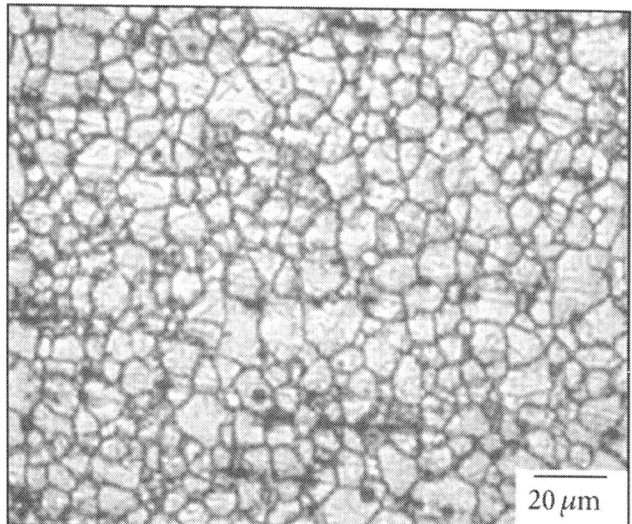

20 μm

Figure 3: Microstructure of friction stir welded stir zone before the corrosion test.

Finding the Limits of Corrosion Test Parameters

From the literature [15, 16], the predominant factors that have a greater influence on corrosion behavior of AZ61A magnesium alloy are identified. They are (i) pH value of the solution, (ii) exposure time, and (iii) chloride ion concentration. Large numbers of trial experiments were conducted to identify the feasible testing conditions using friction stir welded AZ61A magnesium alloy joints under galvanic test conditions. The following inferences are obtained.

- If the pH value of the solution was less than 3 therefore, the change in chloride ion concentration did not considerably affect the corrosion.

- If the pH value was between 3 and 11 therefore, there were inhibition of the corrosion process and stabilization of the protective layer.

- If the pH value was greater than 11 therefore, blocking of further corrosion by the active centers of the partially protective layer.

- If the chloride ion concentration was less than 0.2 M therefore, the visible corrosion did not occur in the experimental period.

- If the chloride ion concentration was between 0.2 M and 1 M therefore, there was a reasonable fluctuation in the corrosion rate.

- If the chloride ion concentration was greater than 1 M therefore, the rise in corrosion rate slightly decreased a little.

- If the exposure time was less than an hour therefore, the surface would be completely covered with thick and rough corrosion products.

- If the exposure time was between 1 and 9 hours therefore, the tracks of the corrosion could be predicted.

- If the exposure time was greater than 9 hours therefore, the tracks of corrosion film were difficult to identify.

Developing the Experimental Design Matrix

Owing to a wide range of factors, the use of three factors and central composite rotatable design matrix was chosen to minimize the number of experiments. The assay conditions for the reaction parameters were taken at zero level (center point) and one level (+1) and (1). The design was extended up to a ±α (axial point) of 1.68. The center values for variables were carried out at least six times for the estimation of error and single runs for each of the other combinations; twenty runs were done in a totally random order. The design would consist of the eight corner points of the 2^3 cube, the six star points, and m center points. The star points would have $\boxtimes = 8^\boxtimes(1/4) = 1.682$. Design matrix consisting of 20 sets of coded conditions (comprising a full replication three factorial of 8 points, six corner points, and six center points) was chosen in this investigation. Table 3 represents the ranges of factors considered, and Table 4 shows the 20 sets of coded and actual values used to conduct the experiments. For the convenience of recording and processing experimental data, the upper and lower levels of the factors were coded here as +1.682 and −1.682, respectively. The coded values of any intermediate values could be calculated using the following relationship:

$$X_i = \frac{1.682\left[2X - \left(X_{max} - X_{min}\right)\right]}{\left(X_{max} - X_{min}\right)},$$

(1)

where $\boxtimes_i$ is the required code values of a variable $\boxtimes$, and $\boxtimes$ is any values of the variable from $\boxtimes_{min}$ to $\boxtimes_{max}$; $\boxtimes_{min}$ is the lower level of the variable, and $\boxtimes_{max}$ is the upper level of the variable.

Table 3: Important factors and their levels

S. no.	Factor	Unit	Notation	Levels				
				−1.682	−1	0	+1	+1.682
1	pH value		P	3	4.62	7	9.38	11
2	Corrosion time	Hours (h)	T	1	2.62	5	7.38	9
3	Cl⁻ concentration	Mole (M)	C	0.2	0.36	0.6	0.84	1

Table 4: Design matrix and experimental results

EX. no.	Coded values			Actual values			Corrosion rate (salt spray tests) (mm/yr)	Corrosion rate (galvanic corrosion tests) (mm/yr)
	pH (P)	Time (T)	Conc. (C)	pH (P)	Time (T)(hour)	Conc. (Mole)		
1	−1	−1	−1	4.62	2.62	0.36	14.62 (0.11)	0.0397 (0.16)
2	+1	−1	−1	9.38	2.62	0.36	10.23 (0.56)	0.0254 (0.13)
3	−1	+1	−1	4.62	7.38	0.36	11.89 (0.21)	0.0340 (0.20)
4	+1	+1	−1	9.38	7.38	0.36	8.99 (0.18)	0.0111 (0.18)
5	−1	−1	+1	4.62	2.62	0.84	15.82 (0.16)	0.0456 (0.14)
6	+1	−1	+1	9.38	2.62	0.84	11.31 (0.1)	0.0425 (0.1)
7	−1	+1	+1	4.62	7.38	0.84	12.92 (0.26)	0.0519 (0.52)
8	+1	+1	+1	9.38	7.38	0.84	10.75 (0.05)	0.0376 (0.05)
9	−1.682	0	0	3	5	0.60	17.96 (0.26)	0.0612 (0.08)
10	+1.682	0	0	11	5	0.60	6.88 (0.41)	0.0267 (0.11)
11	0	−1.682	0	7	1	0.60	11.23 (0.56)	0.0598 (0.07)
12	0	+1.682	0	7	9	0.60	8.51 (0.42)	0.0282 (0.11)
13	0	0	−1.682	7	5	0.20	6.66 (0.23)	0.0184 (0.05)
14	0	0	+1.682	7	5	1	15.28 (0.4)	0.0534 (0.08)
15	0	0	0	7	5	0.60	9.89 (0.36)	0.0432 (0.03)
16	0	0	0	7	5	0.60	8.56 (0.02)	0.0365 (0.02)
17	0	0	0	7	5	0.60	9.79 (0.03)	0.0370 (0.21)
18	0	0	0	7	5	0.60	9.97 (0.05)	0.0377 (0.24)
19	0	0	0	7	5	0.60	8.93 (0.04)	0.0379 (0.38)
20	0	0	0	7	5	0.60	9.62 (0.09)	0.0381 (0.12)

Salt Spray Corrosion Test (Recording the Response)

Solutions of NaCl with concentrations of 0.2 M, 0.361 M, 0.6 M, 0.838 M, and 1 M were prepared. The pH values of the solutions were maintained as pH 3, pH 4.619, pH 7, pH 9.38, and pH 11 with concentrated HCl and NaOH, respectively. The pH value was measured using a digital pH meter. The test method consists of exposing the specimens in a salt spray chamber as per ASTM B 117 standards and evaluating the corrosion tested specimen with the method as per ASTM G1-03. Basically, the salt spray test procedure involves the spraying of a salt solution onto the samples being tested. This was done inside a temperature controlled chamber. The glass racks were contained in the salt fog chamber (3" high, 3" deep, and 5" wide). The samples under test were inserted into the chamber, following which the salt-containing solution was sprayed as a very fine fog mist over the samples. NaCl in tapped water was pumped from a reservoir to spray nozzles. The solution was mixed with humidified compressed air at the nozzle, and this compressed air and atomized the NaCl solution into a fog at the nozzle. Heaters were maintained at 35°C cabinet temperature. Within the chamber, the samples were rotated frequently so that all samples were exposed uniformly to the salt spray mist. Since the spray was continuous, the samples were continuously wet and therefore, uniformly subjected to corrosion. The corrosion rate of the friction stir welded AZ61A alloy specimen was estimated by weight loss measurement. The original weight (w_0) of the specimen was recorded, and then the specimen was sprayed with the solution of NaCl for different spraying times of 1, 2.62, 5, 7.38, and 9 hours. The corrosion products were removed by immersing the specimens for one minute in a solution prepared by using 50 gm chromium trioxide (CrO_3), 2.5 gm silver nitrate ($AgNO_3$), and 5 gm barium nitrate ($Ba(NO_3)_2$) in 250 mL distilled water. Finally, the specimens were washed with distilled water, dried, and weighed again to obtain the final weight (w_1). The weight loss (w) can be measured using the following relation:

$$w = \left(w_o - w_1 \right),$$

(2)

where w is weight loss in grams, w_0 is original weight before test in grams, and w_1 is final weight after test in grams.

The corrosion rate of FSW joints of AZ61A was calculated by using the following equation as per the ASTM standards B117:

$$\text{corrosion rate (mm/year)} = \frac{8.76 \times 10^4 \times w}{A \times D \times T},$$

(3)

where w is weight loss in grams, A is surface area of the specimen in cm^2, D is density of the material (1.72 gm/cm^3), and T is spraying time in hours.

Galvanic Corrosion Test (Recording the Response)

The test method consisted of exposing the specimens in a specially designed apparatus as per ASTM G 82-98 standards and evaluating the corrosion tested specimen with the method as per ASTM G 102-89. The galvanic samples were prepared in the following way. A saturated calomel electrode and graphite electrode were used as the reference and auxiliary electrode, respectively. The working electrodes were the friction stir welded AZ61A magnesium alloy welds coupled with AZ61A magnesium alloy base metal. An electrical contact was made between the galvanic couple, where a Teflon insulation of the same thickness was inserted between the electrodes to avoid the direct contact between the electrodes. The galvanic couple was immersed in NaCl solution with different pH and chloride ion concentration for different immersion times of 1, 2.62, 5, 7.38, and 9 hours. When the mixed potential theory was applied to the individual reactions, the uncoupled corrosion rates were i_{corr} (A) for AZ61A magnesium alloy base metal and i_{corr} (B) for AZ61A friction stir welds. When equal areas of AZ61A Mg base metal and AZ61A friction stir welds were coupled, the resultant mixed potential of the system i_{corr} (AB) was at the intersection where the total oxidation rate equals the total reduction rate. The rate of oxidation of the individual coupled metals was such that the base metal corroded at a reduced rate i_{corr} (A), and AZ61A friction stir welds corroded at an increased rate i_{corr} (B). Hence, the AZ61A friction stir weld acts as an anode, and AZ61A base alloy acts as a cathode. Half-cell reactions were carried out constituting a single cell. Thus, the current i_{corr} (AB) was the galvanic current which can be measured by a zero resistance ammeter (ZRA). Free corrosion potential of both metals was found individually and from the potential difference; FSW AZ61A magnesium alloy was considered to be an anode, because of its more negative potential than AZ61 base alloy, where the latter was the cathode. Corrosion current values may be obtained from the ZRA measurements. The corrosion rate can be calculated using Faraday's Law in terms of penetration rates as follows:

$$\text{corrosion rate (mm/yr)} = \frac{K \times I_{corr} \times EW}{\rho},$$

(4)

where k is corrosion constant (k is 0.00327 if corrosion rate in (mm/yr)), I_{corr} is current density in mA/cm^2, EW is equivalent weight of the alloy, and ρ is density of the FSW AZ61A alloy (1.72 gm/cm^3).

Metallography

Microstructural analysis of the corroded specimens was carried out using a light optical microscope (Union Optics, Japan; model: Versamet-3) incorporated with an image analyzing software (Clemex-vision). The exposed specimen surface was prepared for the microexamination in the "as polished" conditions. The corrosion test specimens were polished in disc-polishing machine with minor polishing, and the surface was observed at 200x magnification. The corrosion products were analyzed by SEM-EDX and XRD analysis.

DEVELOPING AN EMPIRICAL RELA-TIONSHIP

The response surface methodology (RSM) approach was adopted in this study because of its following advantages (1) the ability to evaluate the effects of interactions between tested parameters and (2) the benefit of limiting the number of actual experiments to be carried out, in comparison to a classical approach for the same number of estimated parameters [17–20]. In the present investigation, to correlate the potentiodynamic polarization test parameters and the corrosion rate of AZ61A welds, a second-order quadratic model was developed. The response (corrosion rate of AZ61A welds) is a function of pH values

(P), exposure time (T), and chloride ion concentration (C), and it could be expressed as

$$\text{corrosion rate} = f(P, T, C).$$
(5)

In order to study the combined effects of these parameters, experiments were conducted at different combinations using statistically designed experiments. The empirical relationship must include the main and interaction effects of all factors and hence the selected polynomials are expressed as follows:

$$Y = b_0 + \sum b_i x_i + \sum b_{ii} x_i^2 + \sum b_{ij} x_i x_j.$$
(6)

For three factors, the selected polynomial could be expressed as

$$\begin{aligned}\text{corrosion rate} = \{ &b_0 + b_1(P) + b_2(T) + b_3(C) \\ &+ b_{11}(P^2) + b_{22}(T^2) \\ &+ b_{33}(C^2) + b_{12}(PT) \\ &+ b_{13}(PC) + b_{23}(TC) \},\end{aligned}$$
(7)

where b_0 is the average of responses (corrosion rate), and b_1, b_2, b_3 ... b_{11}, b_{12}, b_{13} ... b_{22}, b_{23}, b_{33} are the coefficients that depend on their respective main and interaction factors, which are calculated using the expression given as follows:

$$B_i = \frac{\sum (X_i, Y_i)}{n},$$
(8)

where "i" varies from 1 to n, in which X_i is the corresponding coded value of a factor, Y_i is the corresponding response output value (corrosion rate) obtained from the experiment, and "n" is the total number of combinations considered. All the coefficients were obtained applying central composite rotatable design matrix including the Design Expert statistical software package. After determining the significant coefficients (at 95% confidence level), the final relationship was developed including only these coefficients. The final empirical relationship obtained by the above procedure to estimate the corrosion rate of friction stir welds of AZ61A magnesium alloy is given as follows.

Salt spray corrosion test,

$$\text{corrosion rate} = \{9.48 - 1.89\,(P) - 0.88\,(T) + 0.82\,(C)$$

$$+1.60\left(P^2\right) + 1.27\left(C^2\right)\}\ \text{mm/yr;} \tag{9}$$

Galvanic corrosion test,

$$\text{corrosion rate} = \{0.056 - 7.33 \times 10^{-3}\,(P)$$

$$+ 2.75 \times 10^{-3}\,(T) + 0.016\,(C)$$

$$- 4.25 \times 10^{-3}\,(PT) + 4.26 \times 10^{-3}\,(PC)$$

$$+ 4.6 \times 10^{-3}\,(TC)$$

$$- 3.74 \times 10^{-4}\left(T^2\right)\}\ \text{mm/yr.} \tag{10}$$

Checking the Adequacy of the Model Salt Spray Testing

The Analysis of Variance (ANOVA) technique was used to find the significant main and interaction factors. The results of the second-order response surface model fitting in the form of Analysis of Variance (ANOVA) are given in Table 5. The determination coefficient (R^2) indicated the goodness of fit for the model. The model F value of 31.30 implies the model is significant. There is only a 0.01% chance that a "model F value" this large could occur due to noise. Values of "Prob > F" less than 0.0500 indicate that model terms are significant. In this case P, T, C, P^2, and C^2 are significant model terms. Values greater than 0.1000 indicate that the model terms are not significant. If there are many insignificant model terms (not counting those required to support hierarchy), model reduction may improve the model. The "lack of fit F value" of 1.69 implies that the lack of fit is not significant relative to the pure error. There is a 28.93% chance that a "lack of fit F value" this large could occur due to noise. No significant lack of fit is good. The "Pred R-Squared" of 0.8176 is in reasonable agreement with the "Adj R-Squared" of 0.9349. "Adeq Precision" measures the signal to noise ratio. A ratio greater than 4 is desirable. Our ratio of 19.440 indicates an adequate signal. All of this indicated an excellent suitability of the regression model. Each of the observed values compared with the experimental values are shown in Figure 4.

Table 5: ANOVA test results for salt spray corrosion test

Source	Sum of squares	df	Mean square	F value	P value	Prob > F
Model	126.60	9	14.07	31.30	<0.0001	Significant
P	49.03	1	49.03	109.11	<0.0001	
T	10.55	1	10.55	23.48	0.0007	
C	9.12	1	9.12	20.29	0.0011	
PT	1.83	1	1.83	4.08	0.0710	
PC	0.047	1	0.047	0.10	0.7543	
TC	0.033	1	0.033	0.072	0.7934	
P^2	37.07	1	37.07	82.48	<0.0001	
T^2	3.4E-004	1	3.4E-004	7.6E-004	0.9785	
C^2	23.17	1	23.17	51.55	<0.0001	
Residual	4.49	10	0.45			
Lack of fit	2.82	5	0.53	1.69	0.2893	Not significant
Pure error	1.67	5	0.33			
Cor total	131.09	19				

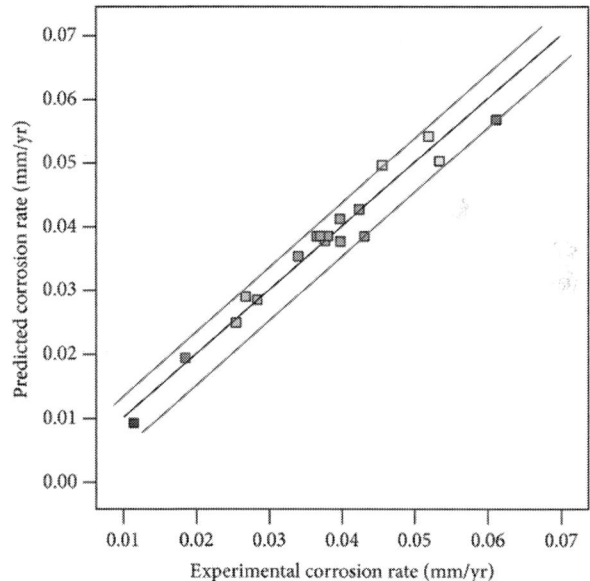

Figure 4: Correlation graph for response (salt spray corrosion test).

Checking the Adequacy of the Model Galvanic Corrosion Testing

The Analysis of Variance (ANOVA) technique was used to find the significant main and interaction factors. The results of second-order response surface model fitting in the form of Analysis of Variance (ANOVA) are given in Table 6. The determination coefficient (R^2) indicated the goodness of fit for the model. The model F value of 27.66 implies the model is significant. There is only a 0.01% chance that a "model F value" this large could occur due to noise. Values of "Prob > F" less than 0.0500 indicate that model terms are significant. In this case A, B, C, AB, and A^2 are significant model terms. Values greater than 0.1000 indicate that the model terms were not significant. If there are many insignificant model terms (not counting those required to support hierarchy), model reduction may improve the model.

Table 6: ANOVA test results for galvanic corrosion test

Source	Sum of squares		df	Mean square		value	value	Prob >
Model	2.472⊠003	—	9	2.472⊠003	—	27.66	<0.0001	Significant
P	9.229⊠004	—	1	9.229⊠004	—	92.93	<0.0001	
T	1.057⊠004	—	1	1.057⊠004	—	10.64	0.0085	
C	1.163⊠004	—	1	1.163⊠004	—	117.12	<0.0001	
PT	4.642⊠005	—	1	4.642⊠004	—	4.67	0.0559	
PC	4.700⊠005	—	1	4.700⊠005	—	4.73	0.0547	
TC	5.634⊠005	—	1	5.634⊠005	—	5.67	0.0385	
P²	2.804⊠005	—	1	2.804⊠005	—	2.82	0.1238	
T²	6.470⊠005	—	1	6.470⊠005	—	6.51	0.0287	

C^2	$3.072 \boxtimes 005$ —	1	$3.072 \boxtimes 005$ —	3.09	0.1091	
Residual	$9.931 \boxtimes 005$ —	10	$9.931 \boxtimes 005$ —			
Lack of fit	$6.999 \boxtimes 005$ —	5	$1.400 \boxtimes 005$ —	2.39	0.1808	Not significant
Pure error	$2.932 \boxtimes 005$ —	5	$5.864 \boxtimes 005$ —			
Cor total	$2.572 \boxtimes 005$ —	19				

The "lack of fit ⊠ value" of 2.39 implies that the lack of fit was not significant relative to the pure error. There was an 18.08% chance that a "lack of fit ⊠ value" this large could occur due to noise. Nonsignificant lack of fit is good. The "Pred ⊠- Squared" of 0.7763 is in reasonable agreement with the "Adj ⊠-Squared" of 0.9266. "Adeq Precision" measures the signal to noise ratio. ⊠ ratio greater than 4 is desirable. Our ratio of 21.393 indicates an adequate signal. Each of the observed values compared with the experimental values are shown in Figure 5, and it had a good agreement between the observed values and the experimental values.

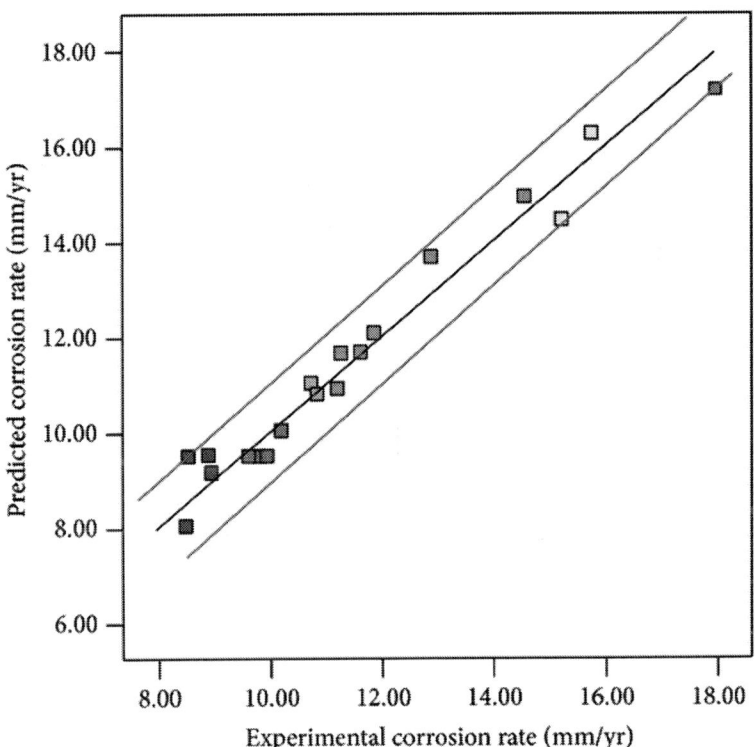

Figure 5: Correlation graph for response (galvanic corrosion test).

RESULTS AND DISCUSSION

Effect of pH on Corrosion Rate

Figure 6 shows the graph representing the effect of pH on corrosion rate during salt spray testing and galvanic corrosion testing. For both corrosion tests, the graph shows clearly that the corrosion rate decreased with the increase in pH value. At every chloride ion concentration and immersion time, the FS welds usually exhibited a decrease in corrosion rate with increase in pH. In neutral pH, the corrosion rate remained approximately constant, and comparatively low corrosion rate was observed in alkaline solution. It was seen that the influence of pH was more at higher concentration as compared to lower concentration in neutral and alkaline solutions.

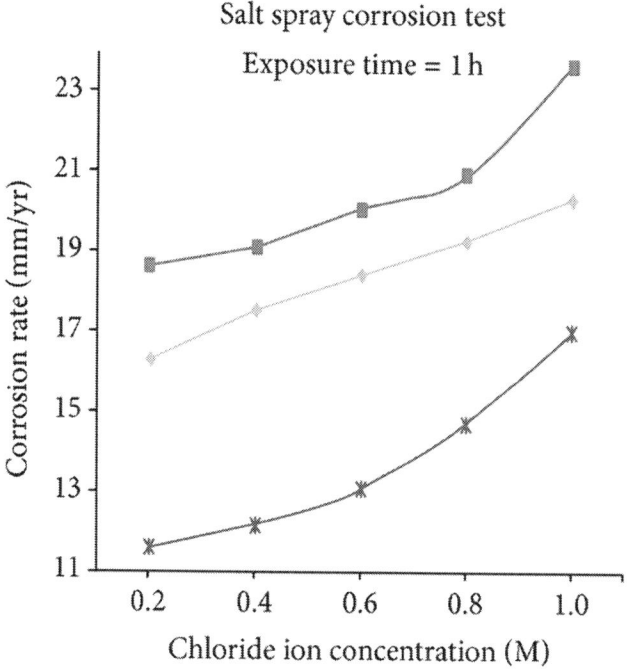

(a)

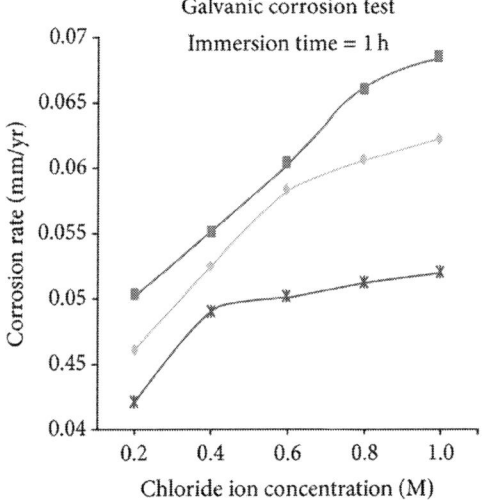

(b)

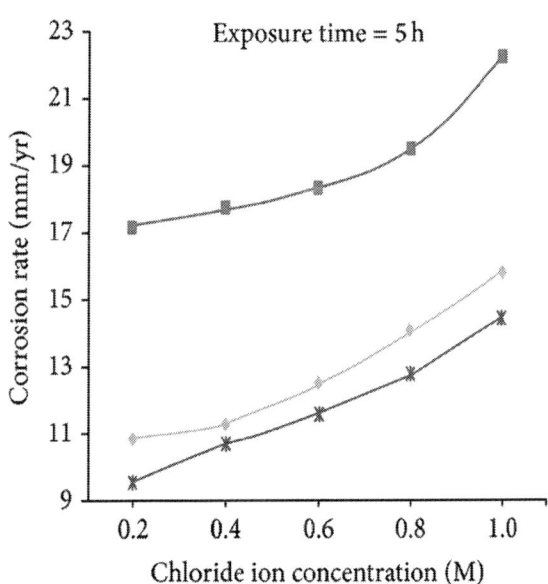

(c)

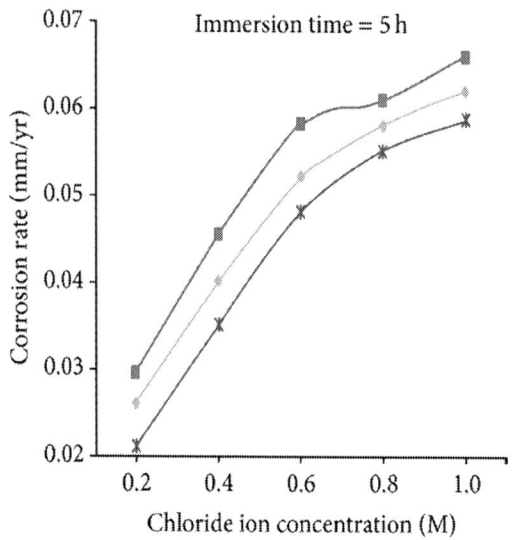

(d)

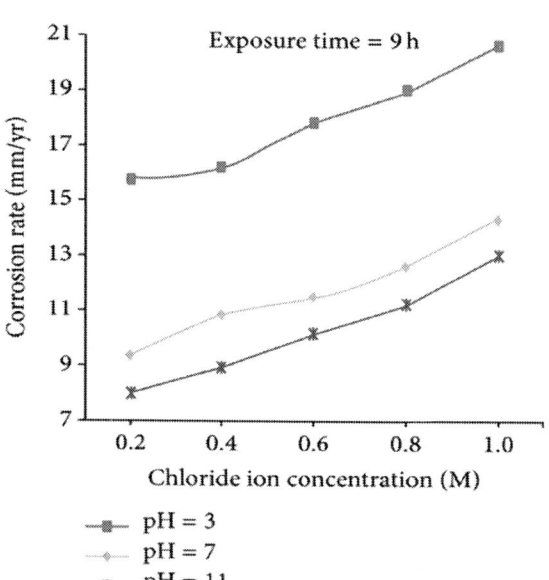

(e)

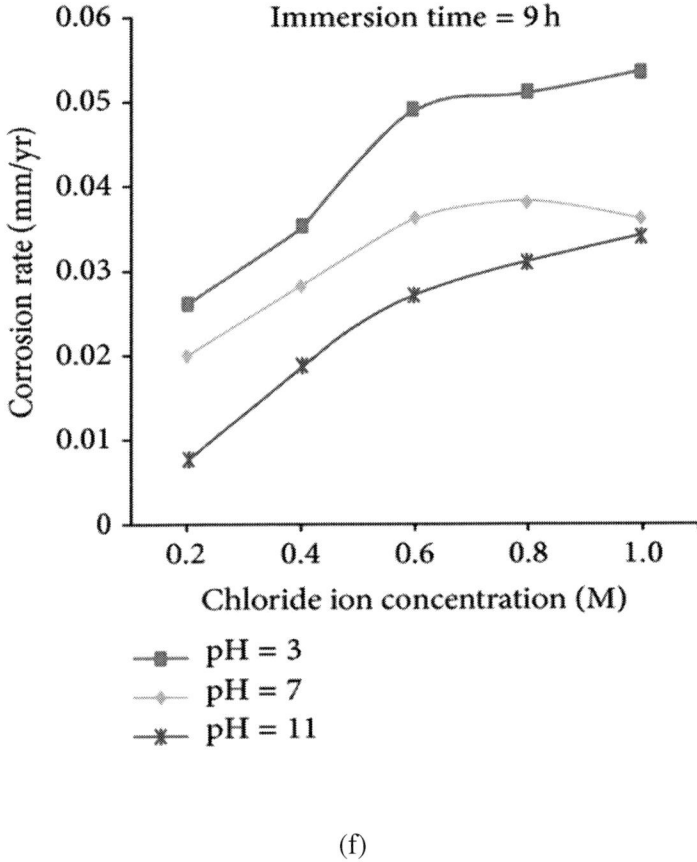

(f)

Figure 6: Effect of pH value on corrosion rate.

On comparing the corrosion rate of both, the corrosion tested specimen was represented as bar diagram in Figure 7. It was found that the corrosion rates obtained from the salt spray testing were much higher than the rates obtained from the galvanic corrosion tests. This was due to spraying effect where recycling of the solution could not be taken into account, while in galvanic corrosion testing; there is a substantial increase in the pH of the solution during immersion testing causing alkalization or basification of the solution with the increase in reactivity and time. Thus, the corrosion rate was much higher in salt spray testing than in the galvanic corrosion testing. So, the couple were galvanically a good couple, and can be suitable for good applications [20–22].

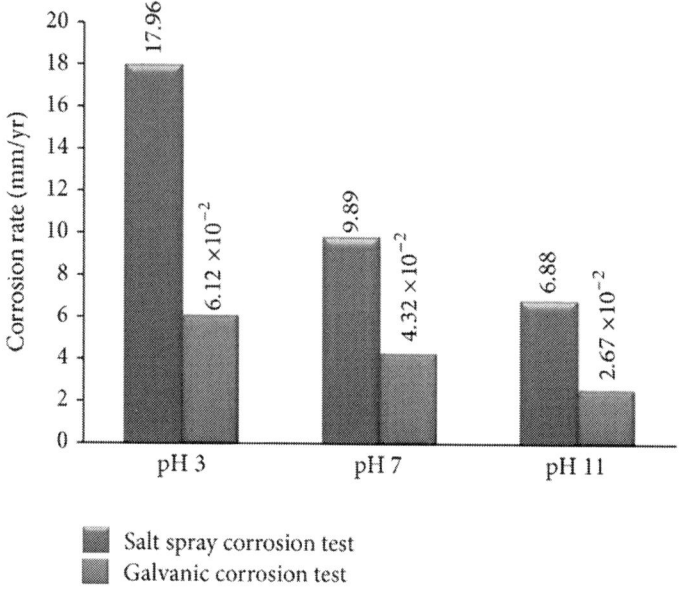

Figure 7: Comparative estimation of corrosion rate with respect to pH value.

Figure 8 shows the effect of pH on pit morphology of the corroded specimen exposed in 0.6 M concentration of NaCl for 5 hours with different pH values of pH 3, pH 7, and pH 11 for both salt spray testing and galvanic corrosion testing. During salt spray testing, the density of the pit formed in exposing lower pH (acidic) solution is quite high, compared with the neutral and alkaline solution. It was observed that the matrix shows the pitting marks and the pitting corrosion that has taken place at the friction stir welded microstructure. The particles are Mn-Al compound and fragmented $Mg_{17}Al_{12}$. The numbers of pits were more in the joints when it is sprayed with the solution of low pH. Hence, the corrosion rate increases with the decrease in pH value. Since the increase of grain and grain boundary in the joints, the grain boundary acts cathodic to grain causing a microgalvanic effect. The presence of microgalvanic effect between the ⊠ phase and the ⊠ phase that formed was due to the presence of aluminum. During galvanic corrosion tests, the grain boundaries of the anodic specimen got attacked, and its gravity varies with the parameters used in the experiment. Corrosion tends to be concentrated in the area adjacent to the grain boundary until eventually the grain may be undercut and fall out [23].

Salt spray corrosion test

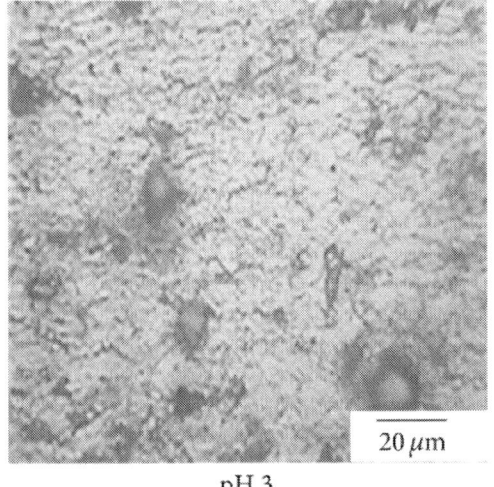

pH 3

(a)

Galvanic corrosion test

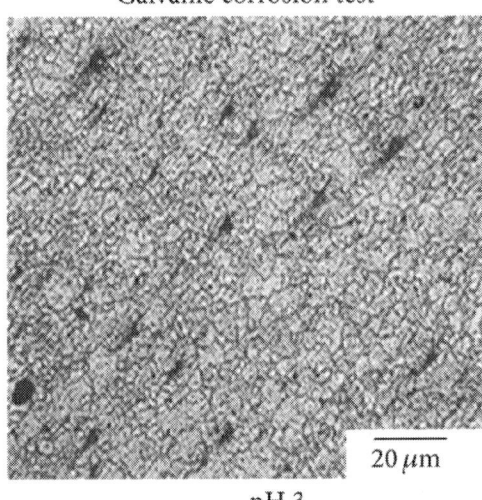

pH 3

(b)

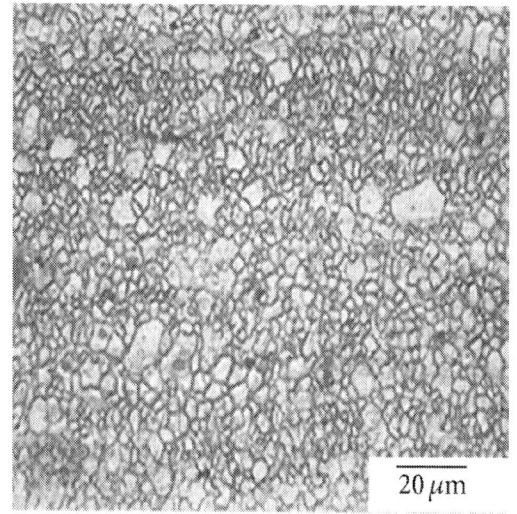

20 µm

pH 7

(c)

20 µm

pH 7

(d)

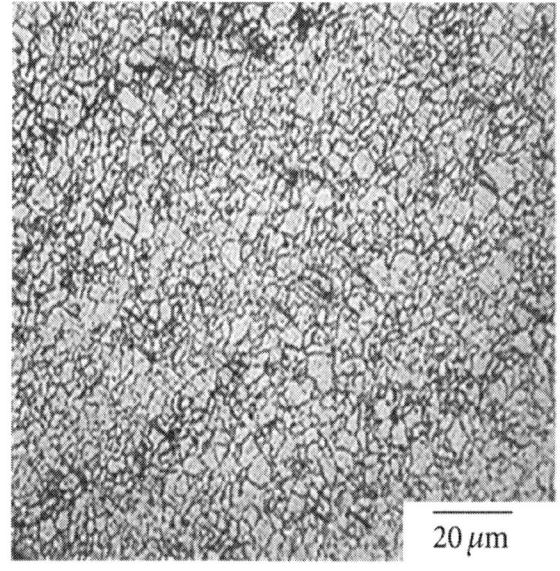

pH 11

(e)

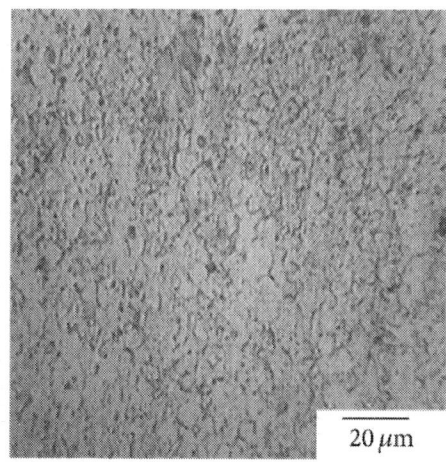

pH 11

(f)

Figure 8: Effect of pH on pit morphology.

Effect of Chloride Ion Concentration on Corrosion Rate

Figure 9 shows the graph representing the effect of chloride ion concentration on corrosion rate during salt spray testing and galvanic corrosion testing. However, it was observed that, with the increase in chloride ion concentration, the rising rate of corrosion rate decreased. The increase in corrosion rate with increasing chloride ion concentration may be attributed to the participation of chloride ions in the dissolution reaction for both corrosion tests [24]. Figure 10 represents the comparison chart for the corrosion rate obtained from both corrosion tests. This is consistent with the detailing of the protective layer. With the increase of chloride ion concentration, the protective layer $Mg(OH)_2$ changed into soluble $MgCl_2$ layer in salt spray corrosion and $Mg(OH)Cl_2$ in galvanic corrosion. The corrosion rate was quite higher in salt spray corrosion test than in the galvanic corrosion test. It states that the $MgCl_2$ was highly soluble compared to $Mg(OH)Cl_2$.

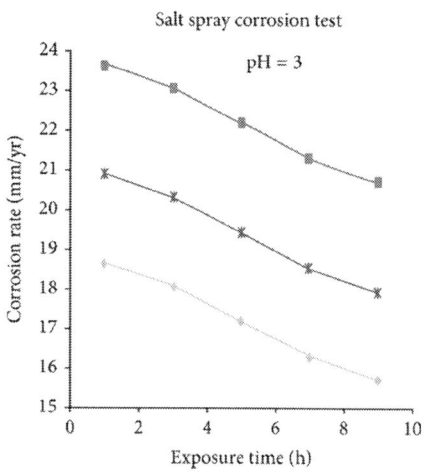

(a)

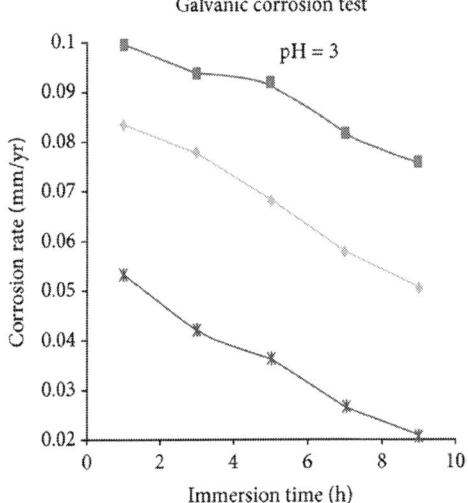

(b)

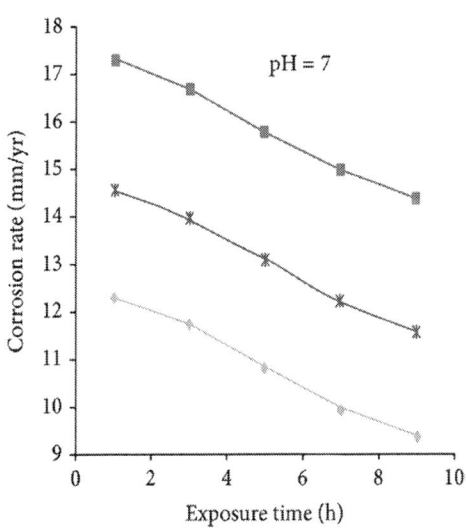

(c)

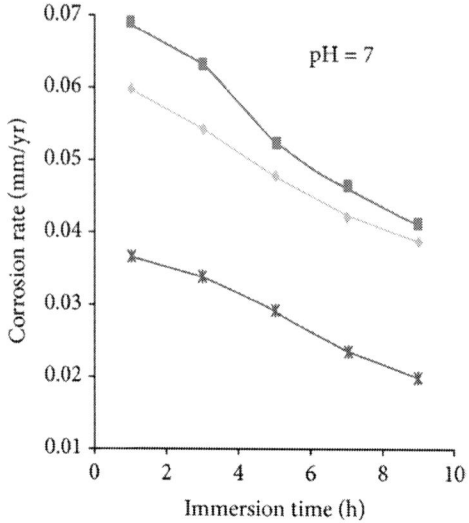

(d)

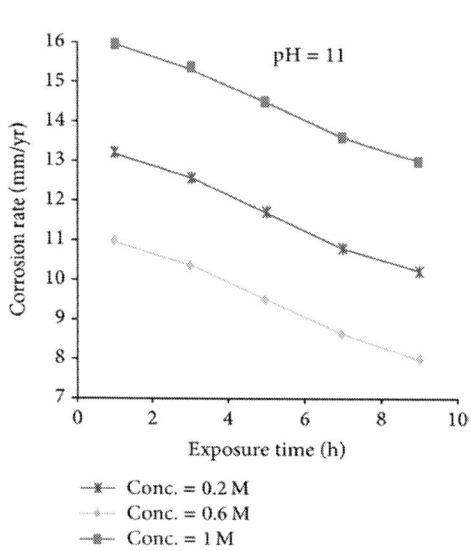

(e)

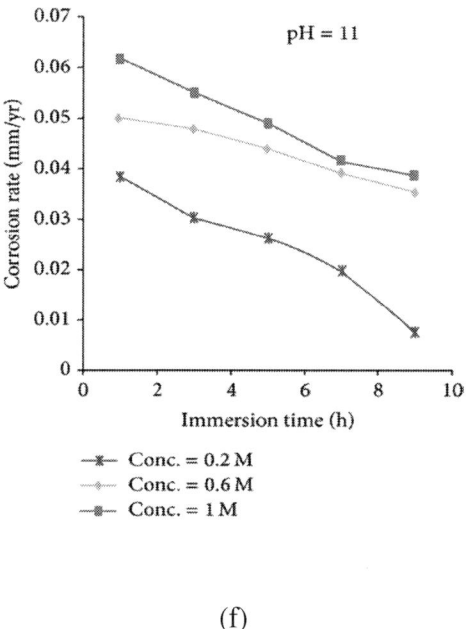

(f)

Figure 9: Effect of chloride ion concentration on corrosion rate.

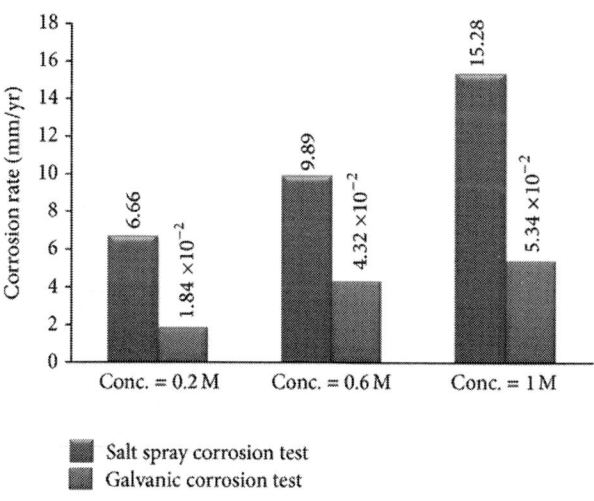

Figure 10: Comparative estimation of corrosion rate with respect to chloride ion concentration.

Figure 11 shows the effect of chloride ion concentration on pit morphology of the corroded specimen exposed in pH 7 for 5 hours with different chloride ion concentration of 0.2 M, 0.6 M, and 1 M for both salt spray testing and galvanic corrosion testing. During salt spray testing, it showed that the alloy exhibited a rise in corrosion rate with the increase in Cl$^-$ concentration and thus, the change of Cl$^-$ concentration affected the corrosion rate much more in higher concentration solutions than that in lower concentration solutions. When more Cl$^-$ in NaCl solution promoted the corrosion, the corrosive intermediate (Cl$^-$) would be rapidly transferred through the outer layer and reache the substrate of the alloy surface. Hence, the corrosion rate increased [20]. But in galvanic corrosion tests, the anodic specimen exhibited a rise in corrosion rate with increase in Cl$^-$ concentration and thus, the change of Cl$^-$ concentration affected the corrosion rate much more in higher concentration solutions than that in lower concentration solutions. Chloride ions were aggressive for magnesium. The adsorption of chloride ions to oxide covered magnesium surface transformed $Mg(OH)_2$ to easily soluble $MgCl_2$. It was considered that the corrosion becomes severe owing to the penetration of the hydroxide film by Cl$^-$ ion and thereby caused the formation of metal hydroxyl chloride complex which governed the following reaction:

$$Mg^{2+} + 2H_2O + 2Cl^- \longrightarrow 2Mg(OH)_2Cl_2$$

$$(11)$$

Salt spray corrosion test

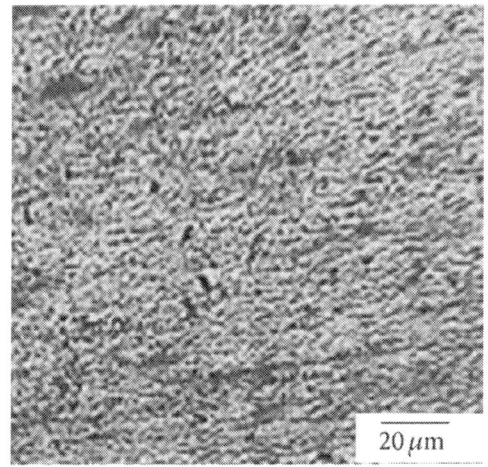

Chloride ion concentration = 0.2 M

(a)

Galvanic corrosion test

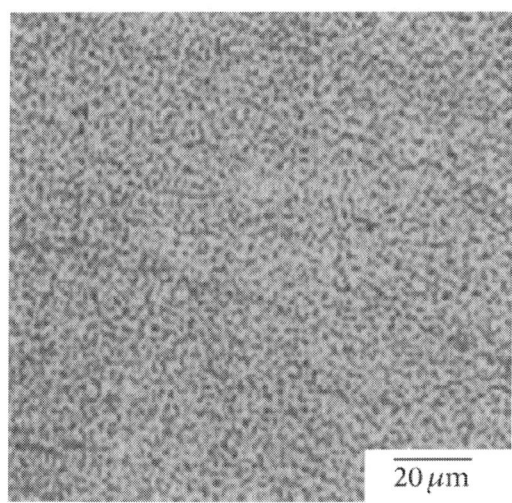

Chloride ion concentration = 0.2 M

(b)

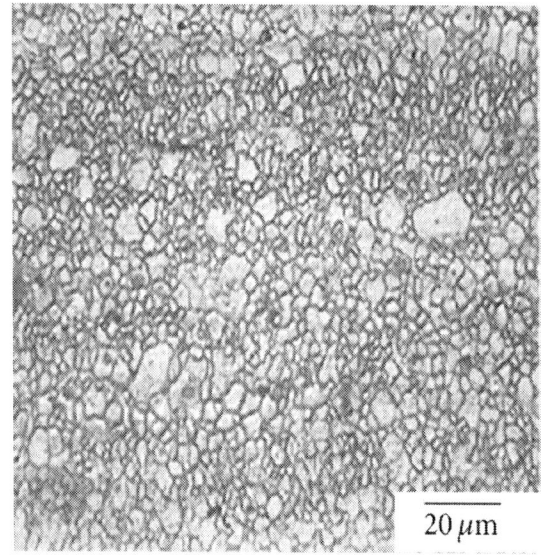

Chloride ion concentration = 0.6 M

(c)

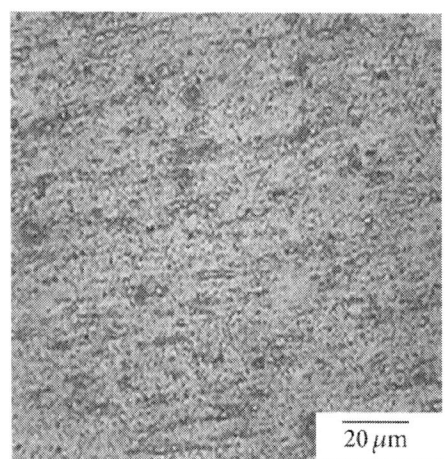

Chloride ion concentration = 0.6 M

(d)

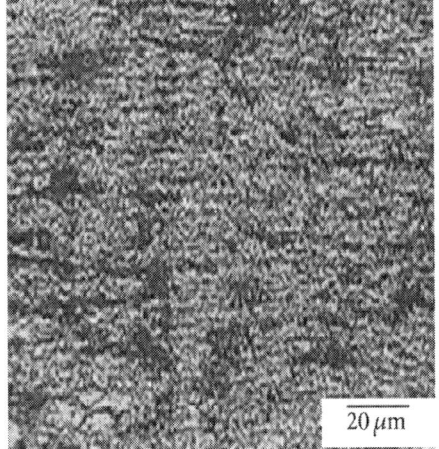

Chloride ion concentration = 1 M

(e)

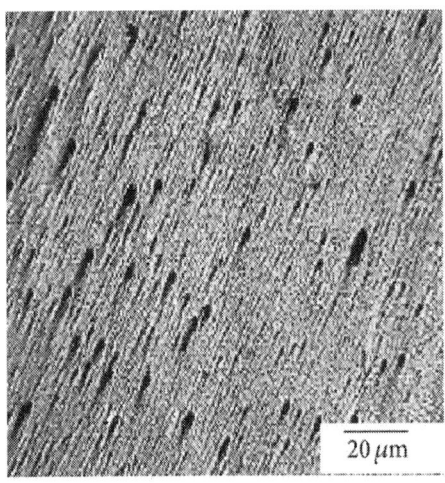

Chloride ion concentration = 1 M

(f)

Figure 11: Effect of chloride ion concentration on pit morphology.

Effect of Corrosion Time on Corrosion Rate

Figure 12 shows the graph representing the effect of corrosion time on corrosion rate during salt spray testing and galvanic corrosion testing. During salt spray testing, the graph shows clearly that the corrosion rate was decreased with the increase in exposure time. It resulted in an increase in hydrogen evolution with the increasing exposure time, which tends to increase the concentration of OH⁻ ions strengthening the surface from causing further corrosion. Thus, the rate of corrosion decreases with the increase in corrosion time. During galvanic corrosion testing, the corrosion rate decreases with the increase in immersion time. The increase in immersion time enhanced the tendency to form the corrosion products, which accumulated over the surface of the samples. These corrosion products in turn depressed the corrosion rate due to the passivation in the medium immersion [25]. It resulted in an increase in hydrogen evolution with the increasing immersion time, which tends to increase the concentration of OH⁻ ions strengthening the surface from causing further corrosion. The strength of the electrolyte reduces from acidity to alkalinity with the increase of time.

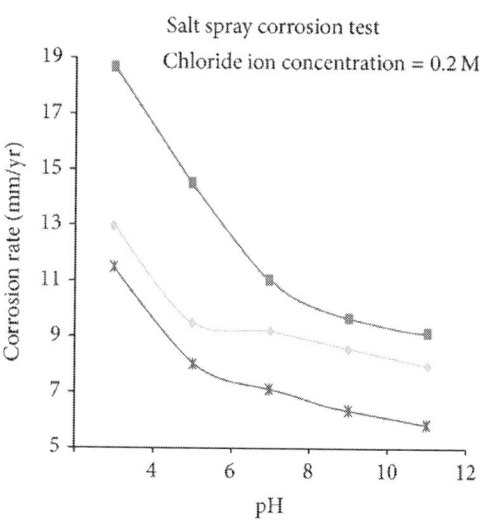

(a)

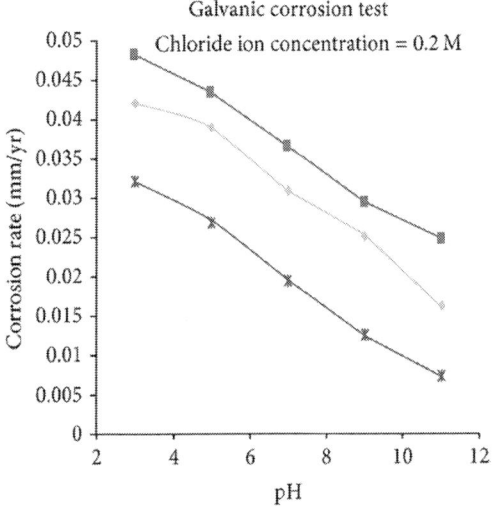

(b)

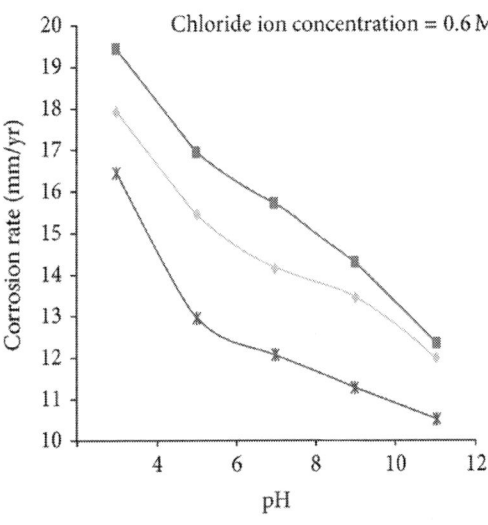

(c)

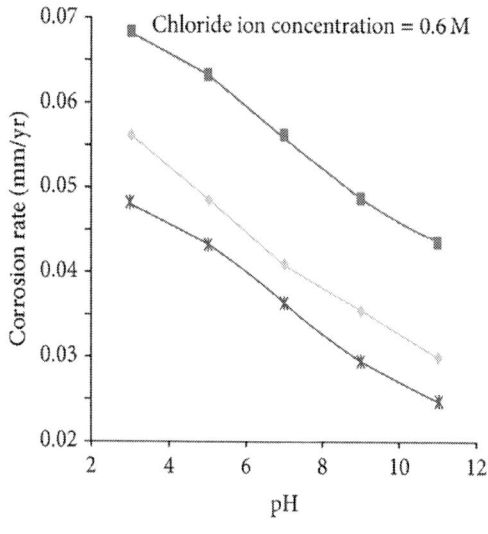

(d)

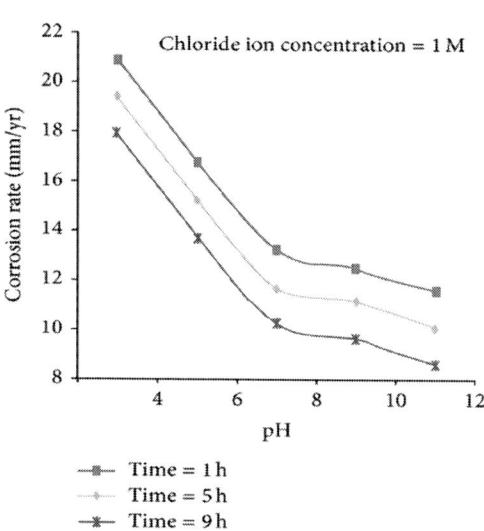

(e)

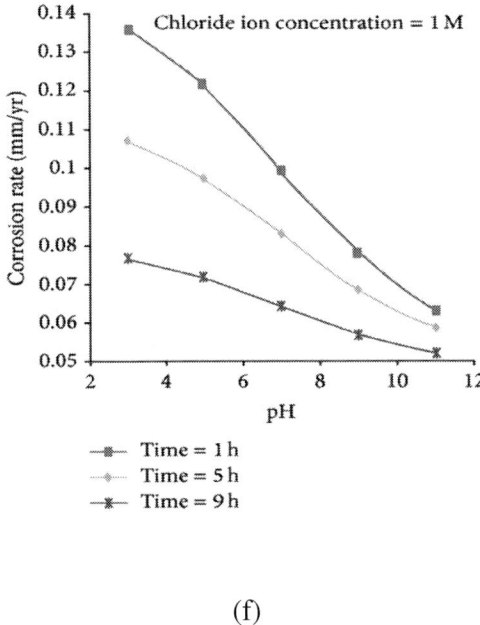

(f)

Figure 12: Effect of corrosion time on corrosion rate.

This is attributed to corrosion occurs over an increasing fraction on the surface leaving the white flakes, which is the insoluble corrosion product [24]. The insoluble corrosion products on the surface of the alloy could slow down the corrosion rate:

$$Mg \longrightarrow Mg^{2+} + 2e^- \tag{12}$$

$$2H_2O + 2e^- \longrightarrow 2OH^- + H_2 \tag{13}$$

$$Mg^{2+} + 2OH^- \longrightarrow Mg(OH)_2 \tag{14}$$

Figure 13 shows the comparison of the corrosion rate obtained during the salt spray and galvanic corrosion test. With the increase of corrosion time the corrosion rate decreases for both specimens. It proved that the protective layer made a predominant role to strike against corrosion with the increment of time. The corrosion rate seems higher in salt spray corrosion test due to the spraying effect, while in immersed condition, the protective layer formed during galvanic corrosion was enhanced by the alkalization of the solution.

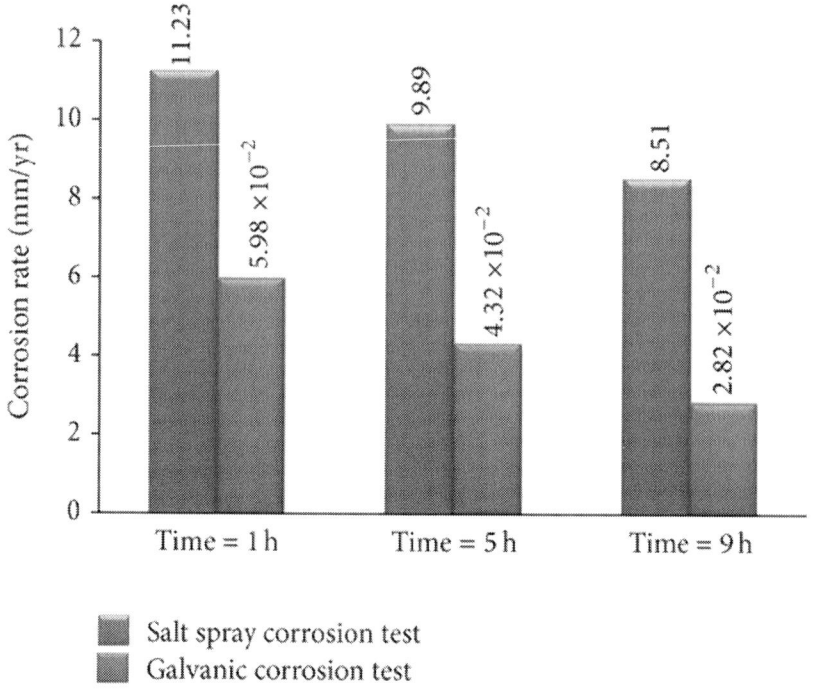

Figure 13: Comparative estimation of corrosion rate with respect to corrosion time.

Figure 14 shows the effect of corrosion time on pit morphology of the corroded specimen exposed in pH 7 with chloride ion concentration of 0.6 M NaCl exposed, 1 h, 5 h, and 9 h for both salt spray testing and galvanic corrosion testing. The mode of microstructural features was comparatively the same during corrosion testing for both tests as its corrosion time is taken into account. The FS welded specimens possess refined grain, and quite a lot of ⬚ particles were distributed continually along the grain boundary.

Salt spray corrosion test

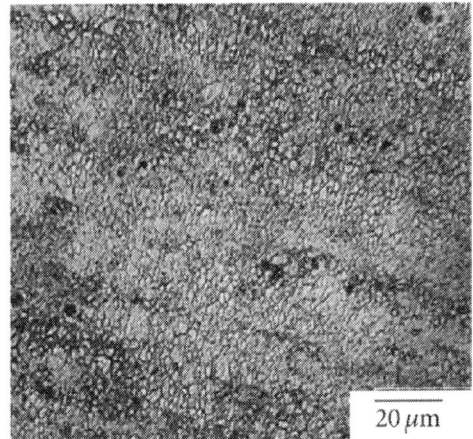

20 μm

Corrosion time = 1 h

(a)

Galvanic corrosion test

20 μm

Corrosion time = 1 h

(b)

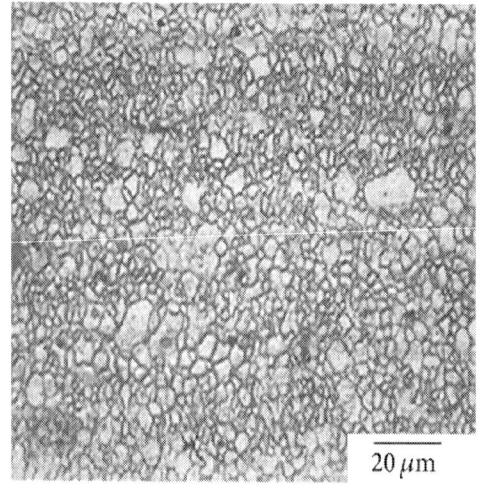

20 µm

Corrosion time = 5 h

(c)

20 µm

Corrosion time = 5 h

(d)

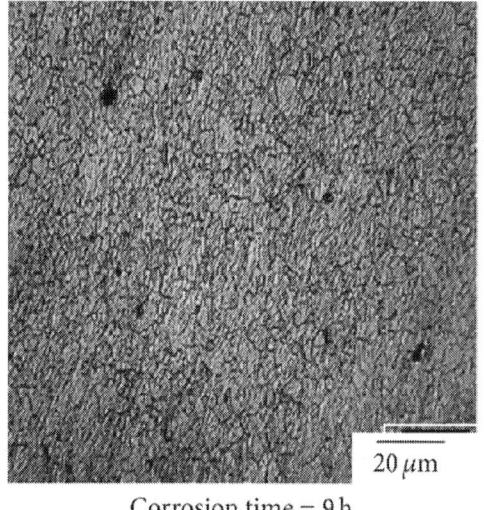

20 μm

Corrosion time = 9 h

(e)

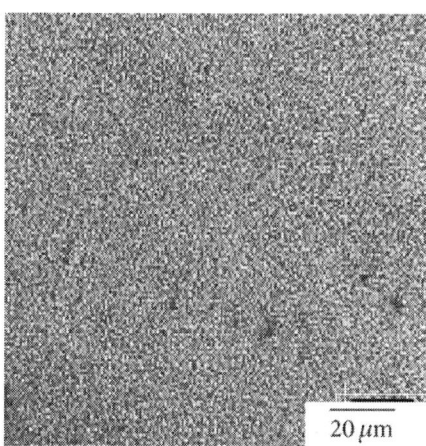

20 μm

Corrosion time = 9 h

(f)

Figure 14: Effect of corrosion time on pit morphology.

In this case, β phase particles cannot be easily destroyed and, with the increase of corrosion time, the quantity of β phases in the exposed surface would increase and finally play the role of corrosion barrier [26]. Although there are some grains of α phase still being corroded, most of the remaining α phase grains are protected under the β phase barrier, so the corrosion rate decreased with the increase in corrosion time. Thus, the corrosion morphology of the alloy was predominantly controlled by the β phase distribution [27].

SEM and XRD Analysis

Figure 15 shows the surface texture of the specimens that underwent salt spray corrosion and galvanic corrosion tests which were observed under SEM as corrosion time was a factor. Figure 15(a) shows the specimens exposed to 1 hour comprised of more localized attack. With corrosion time as a factor, the less spraying time tends to attack more locally on the surface, and later, it penetrates to the substrates, causing higher corrosion behavior and corrosion rate. It was redeems to spalling of corrosion products. Thus, serious pitting occurred in the surface of the weldment with less exposure time for both corrosion-tested specimens. The quantity of corrosion products formed was quite comparably larger in the galvanic corrosion test caused due to immersion.

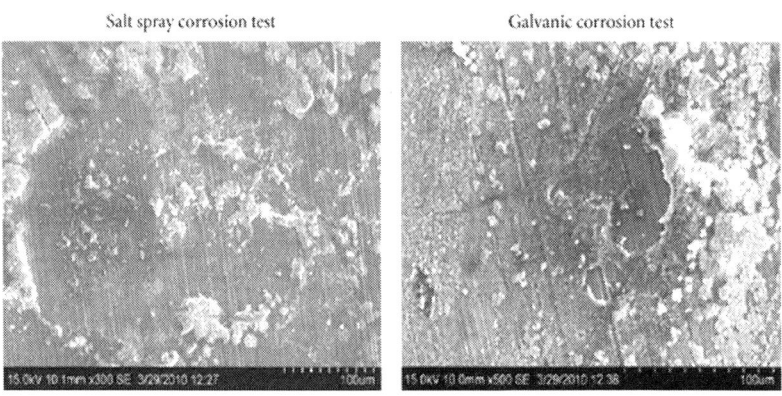

(a)

(b)

Figure 15: Scanning electron micrograph of corrosion test specimens underwent salt spray corrosion test and galvanic corrosion test.

Figure 15(b) shows the specimens exposed to 9 hours composed of more corrosion products. With the increase in corrosion time, the hydroxide layer formed is the dominant factor to avoid further corrosion. This is attributed to corrosion occurring over an increasing fraction of the surface, which is the insoluble corrosion product, $Mg(OH)_2$. Thus, the corrosion rate decreases with the increase of corrosion time. It was observed that the corrosion products were thick and adherent in the specimen that underwent galvanic corrosion test while the specimen that underwent salt spray test exhibited lamellar corrosion products, which seems higher attack and less protective. This showed higher corrosion rate in the salt spray test than the galvanic corrosion test.

Figure 16 shows the XRD analysis to predict the composition of corrosion products and phase in the specimen subjected to salt spray and galvanic corrosion tests. Figure 16(a) shows that the specimen underwent the salt spray corrosion test; the characteristic peaks originate from the metallic Mg substrate and the β phase. The detected number of peaks relates to the β phase ($Mg_{17}Al_{12}$) which is higher in its intensities. It symbolizes the hydroxide layer which finds it hard to form during the salt spray corrosion test due to its spraying effects. However, $Mg(OH)_2$ and MgO phases are detected in Figure 16(b), where the specimen underwent the galvanic corrosion test; $Mg(OH)_2$ is the dominant product in the corrosion zone of the anodic specimen that underwent the galvanic corrosion test. $Mg(OH)_2$ (brucite) has a hexagonal crystal structure and easily undergoes basal cleavage causing cracking and curling in the film, which could play a major role in reducing the corrosion behavior and the corrosion rate. It signifies during immersion that the alkalization effect tends to strengthen the formation of the hydroxide layer.

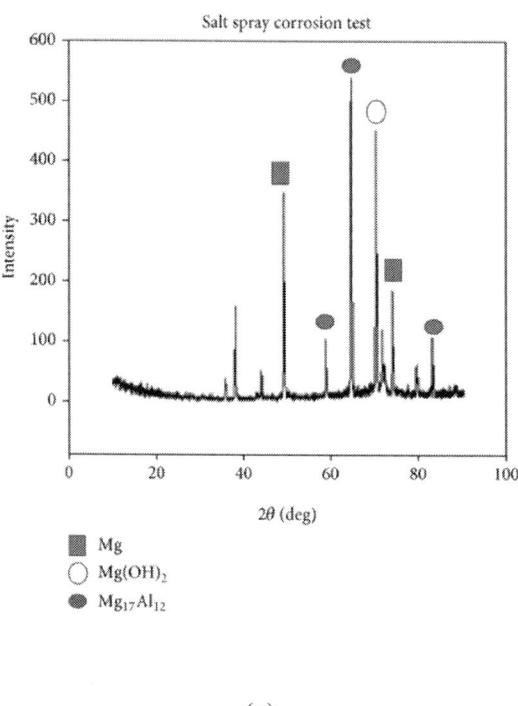

(a)

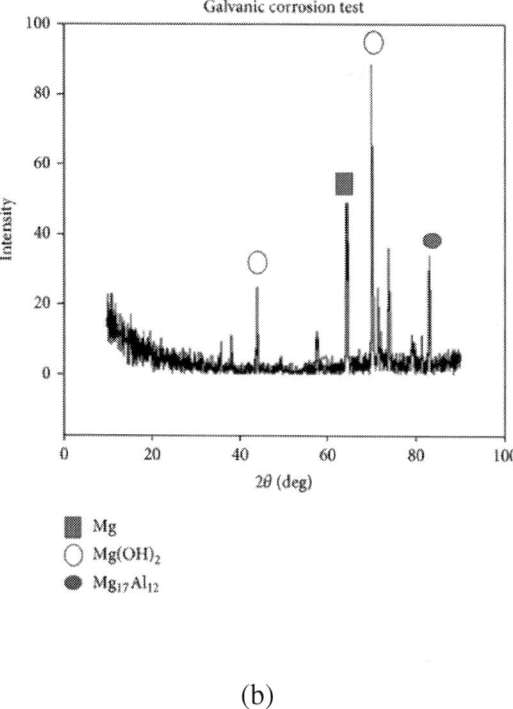

(b)

Figure 16: XRD pattern of corrosion test specimens underwent salt spray corrosion test and galvanic corrosion test.

CONCLUSIONS

1. A mathematical model has been developed here to predict the corrosion rate using salt spray tests and galvanic corrosion with a 95% confidence level. The corrosion rates obtained were quite different in both tests; it found that, the corrosion rate was much higher in salt spray test than in the galvanic corrosion test.

2. In this investigation, it was proved from both tests, for every pH value; that the FS weld metal exhibited a rise in corrosion rate with decrease in pH value. In the neutral pH, the corrosion rate remained approximately constant in neutral solutions, and a comparatively low corrosion rate was observed in alkaline solutions.

3. Chloride ions were aggressive on magnesium alloy. The adsorption of chloride ions to oxide covered the magnesium surface and transformed $Mg(OH)_2$ to easily soluble $MgCl_2$ in salt spray corrosion testing while in galvanic corrosion testing, it readily formed $Mg(OH)_2Cl_2$ due to the immersion criterion.

4. It resulted in an increase in hydrogen evolution with the increasing corrosion time, which tended to increase the concentration of OH^- ions; thereby, an increasing fraction of the surface was observed, which is the insoluble corrosion products. The insoluble corrosion products on the surface of the alloy could slow down the corrosion rate.

5. In this investigation, it was found that the corrosion rate obtained from the salt spray tests was much higher than the rates obtained from the galvanic corrosion tests. It was due to spraying effect where stagnation of the solution could not be taken into account while in galvanic corrosion testing, there is a substantial increase in the pH of the solution during corrosion reactions due to the migration of ions in the electrolyte under immersion. The corrosion rate of galvanic couple ranges from 0.03 to 0.06 mm/yr, which is quite negligible, and shows excellent property of corrosion resistant as per corrosion handbooks and guides. So, the couple were galvanically a good couple, and they can be suitable for good applications.

ACKNOWLEDGMENTS

The authors would like to thank the Centre for Materials Joining & Research (CEMAJOR), Department of Manufacturing Engineering, Annamalai University, Annamalai Nagar, India for extending the facilities of Materials Joining Laboratory and Corrosion Testing Laboratory to carry out this investigation.

REFERENCES

6. B. L. Mordike and T. Ebert, "Magnesium Properties, applications, potential," Materials Science and Engineering A, vol. 302, no. 1, pp. 37–45, 2001. ··

7.   R. C. Zeng, W. Dietzel, R. Zettler, J. Chen, and K. U. Kainer, "Microstructure evolution and tensile properties of friction-stir-welded AM50 magnesium alloy," Transactions of Nonferrous Metals Society of China, vol. 18, no. 1, pp. s76–s80, 2008. ··

8.   R. C. Zeng, J. Zhang, W. J. Huang et al., "Review of studies on corrosion of magnesium alloys,"Transactions of Nonferrous Metals Society of China, vol. 16, supplement 2, pp. s763–s771, 2006. ·

9.   T. Nagasawa, M. Otsuka, T. Yokota, and T. Ueki, "Structure and mechanical properties of friction stir weld Joints of magnesium alloy AZ31," in Magnesium Technology 2000, H. I. Kaplan, J. Hryn, and B. Clow, Eds., pp. 383–387, TMS, Warrendale, Pa, USA, 2000.

10. W. Xu, J. Liu, and H. Zhu, "Pitting corrosion of friction stir welded aluminum alloy thick plate in alkaline chloride solution," Electrochimica Acta, vol. 55, no. 8, pp. 2918–2923, 2010. ··

11. M. Zhao, S. Wu, J. R. Luo, Y. Fukuda, and H. Nakae, "A chromium-free conversion coating of magnesium alloy by a phosphate-permanganate solution," Surface and Coatings Technology, vol. 200, no. 18-19, pp. 5407–5412, 2006. ··

12. B. A. Shaw, "Corrosion resistance of magnesium alloys," in ASM Handbook, vol. 13A: Corrosion, L. J. Korb, Ed., p. 692, ASM International Handbook Committee, Metals Park, Ohio, USA, 9th edition, 2003.

13. D. L. Hawke, J. E. Hillis, M. pekguleryuz, and I. Nkatusugawa, "Corrosion behavior," in Magnesium and Magnesium Alloys, M. M. Avedesian and H. Baker, Eds., pp. 194–1210, ASM International, Materials Park, Ohio, USA, 1999.

14. G. Song and A. Atrens, "Recent insights into the mechanism of magnesium corrosion and research suggestions," Advanced Engineering Materials, vol. 9, no. 3, pp. 177–183, 2007. ·

15. G. Song, B. Johanesson, S. Hagupoda, and D. StJohn, "Galvanic corrosion of magnesium alloy AZ91D in contact with an aluminium alloy, steel and zinc," Corrosion Science, vol. 46, no. 4, pp. 955–977, 2004. ·

16. M. Jönsson, D. Persson, and D. Thierry, "Corrosion product formation during NaCl induced atmospheric corrosion of

magnesium alloy AZ91D," Corrosion Science, vol. 49, no. 3, pp. 1540–1558, 2007. ··

17. R. G. Song, C. Blawert, W. Dietzel, and A. Atrens, "A study on stress corrosion cracking and hydrogen embrittlement of AZ31 magnesium alloy," Materials Science and Engineering A, vol. 399, no. 1-2, pp. 308–317, 2005. ··

18. M. B. Kannan, W. Dietzel, C. Blawert, S. Riekehr, and M. Koçak, "Stress corrosion cracking behavior of Nd:YAG laser butt welded AZ31 Mg sheet," Materials Science and Engineering A, vol. 444, no. 1-2, pp. 220–226, 2007. ··

19. R. Baboian, "Electrochemical techniques for corrosion engineering," in Corrosion ‹76, p. 114, NACE, 1976.

20. H. Altun and S. Sen, "Studies on the influence of chloride ion concentration and pH on the corrosion and electrochemical behaviour of AZ63 magnesium alloy," Materials and Design, vol. 25, no. 7, pp. 637–643, 2004. ··

21. Y. Song, D. Shan, R. Chen, and E. H. Han, "Effect of second phases on the corrosion behaviour of wrought Mg-Zn-Y-Zr alloy," Corrosion Science, vol. 52, no. 5, pp. 1830–1837, 2010. ··

22. K. H. Goh, T. T. Lim, and P. C. Chui, "Evaluation of the effect of dosage, pH and contact time on high-dose phosphate inhibition for copper corrosion control using response surface methodology (RSM),"Corrosion Science, vol. 50, no. 4, pp. 918–927, 2008. ··

23. N. Aslan, "Application of response surface methodology and central composite rotatable design for modeling and optimization of a multi-gravity separator for chromite concentration," Powder Technology, vol. 185, no. 1, pp. 80–86, 2008. ··

24. J. S. Cowpe, J. S. Astin, R. D. Pilkington, and A. E. Hill, "Application of response surface methodology to laser-induced breakdown spectroscopy: Influences of hardware configuration," Spectrochimica Acta B, vol. 62, no. 12, pp. 1335–1342, 2007. ··

25. A. Dhanapal, S. R. Boopathy, and V. Balasubramanian, "Developing an empirical relationship to predict the corrosion rate of friction stir welded AZ61A magnesium alloy under salt fog environment,"Materials and Design, vol. 32, no. 10, pp. 5066–5072, 2011. ··

26. B. D. Craig and D. B. Anderson, Handbook of Corrosion Data, ASM International, 1995.

27. H. H. Uhlig, The Corrosion Handbook, John Wiley, 1948.

28. N. Hara, Y. Kobayashi, D. Kagaya, and N. Akao, "Formation and breakdown of surface films on magnesium and its alloys in aqueous solutions," Corrosion Science, vol. 49, no. 1, pp. 166–175, 2007. · ·

29. G. Song, A. Atrens, and M. Dargusch, "Influence of microstructure on the corrosion of diecast AZ91D,"Corrosion Science, vol. 41, no. 2, pp. 249–273, 1998. · ·

30. Z. M. Zhang, H. Y. Xu, and B. C. Li, "Corrosion properties of plastically deformed AZ80 magnesium alloy," Transactions of Nonferrous Metals Society of China, vol. 20, no. 2, pp. s697–s702, 2010. · ·

31. Y. Song, D. Shan, R. Chen, and E. H. Han, "Effect of second phases on the corrosion behaviour of wrought Mg-Zn-Y-Zr alloy," Corrosion Science, vol. 52, no. 5, pp. 1830–1837, 2010. · ·

32. Corrosion Tests and Standards: Application and Interpretation, ASTM international, 2005.

Chapter 3

One-Step Anodization/Sol-Gel Deposition of Ce3+-Doped Silica-Zirconia Self-Healing Coating on Aluminum

N. Kumar, A. Jyothirmayi, K. R. C. Soma Raju, V. Uma, and R. Subasri

International Advanced Research Centre for Powder Metallurgy and New Materials (ARCI), Balapur, Hyderabad, Andhra Pradesh 500005, India

ABSTRACT

A novel process was used for the preparation of dense, thick, and stable silica-zirconia coatings on aluminum by an in situ anodization along with sol-gel deposition. Anodic electrophoretic deposition was carried on aluminum using a SiO_2-ZrO_2 sol that was synthesized from an epoxy modified silane and zirconium n-propoxide along with a cerium salt ($Ce(NO_3)_3 \cdot 6H_2O$). Current density and time were varied during the deposition. The optimal parameters that yielded uniform coatings were determined. Coatings were characterized for their crystallinity, scratch hardness, and microstructure. The barrier properties of the coatings were tested using potentiodynamic polarization studies, electrochemical impedance spectroscopy, and neutral salt spray tests. Grazing angle incidence X-ray diffraction studies revealed that the coating comprised crystalline Al_2SiO_5 along with an amorphous phase. The novelty of the process was that the crystalline aluminosilicate phase was formed even at room temperature and could be deposited on aluminum by a simultaneous anodization of aluminum and sol-gel deposition. The coated substrates withstood more than 400 hours of salt spray tests. Polarization measurements reveal that the composite layer of aluminosilicate along with the Ce^{3+}-doped silica-zirconia sol enhances the corrosion properties by forming a passive layer, which acts as a good barrier against corrosion.

INTRODUCTION

Aluminum, that is widely used as a structural material due to its high strength to weight ratio and low cost, is highly susceptible to corrosion attack in chloride containing environment [1, 2]. Chromate conversion coatings have been the most widely used self-healing, anticorrosion treatments for aluminum and its alloys. Due to the increasing demand for the development of an environmentally-friendly, effective, inexpensive, and technologically simple method for corrosion protection, anodization of aluminum, and conversion coatings based on vanadates, permanganates, tungstates, and rare-earth salts have recently attracted a lot of attention as a method for corrosion protection [3]. Anodization is an electrochemical oxidation process employed to increase the thickness of the native oxide layer on metals like Al, Mg, Ti, and so forth [4–15]. The anodized layer is porous and, hence, there are reports in which a sol-gel coating is deposited on top of the anodized layer to render good barrier properties [16, 17]. There are also reports where, initially, a purely inorganic sol-gel coating is deposited on the aluminum substrate, followed by heat treatment at high temperatures (300–500°C) for densification, and then the substrate is subsequently anodized [18, 19] in an effort to obtain the cumulative benefits of both processes, namely, anodization and sol-gel coating. However, a low processing temperature for coating densification is one of the most important requisites, since some metals like aluminum undergo structural changes with temperature which degrade the mechanical properties of the substrate and promote corrosion [20]. The sol-gel organic-inorganic hybrid coating is a promising alternative that offers several advantages such as low temperature densification, good adhesion with the substrate, cost-effectiveness, being eco-friendly, and simple application procedures, which are easily adaptable by the user-industry. The sol-gel hybrid matrices can also be used for encapsulating self-healing materials [21, 22].

The present work describes a novel process where aluminum substrate is dipped in a silica-zirconia hybrid sol containing Ce^{3+} which acts as a self-healing material and anodic electrodeposition is carried out. Under these conditions, aluminum can be simultaneously anodized forming, in situ, a crystalline aluminosilicate layer on the substrate surface, along with the deposition of sol-gel coating on the crystalline aluminosilicate layer. The pH of the sol, current density, deposition time, and curing temperature were optimized to form a defect-free, dense, and corrosion resistant oxide layer.

EXPERIMENTAL

Materials and Synthesis

3-Glycidoxypropyltrimethoxysilane (GPTMS, Gelest Inc., USA, purity > 97%), zirconium n-propoxide (GELEST Inc., USA), methacrylic acid (ABCR GmbH, Germany, purity > 97%), and high purity solvent 2-butoxyethanol (LR grade) (Sd Fine-Chem. Ltd., India) were used as the starting materials without further purification. Cerium nitrate $(Ce(NO_3)_3 \cdot 6H_2O)$ 99.9% (LOBA CHEMIE Pvt., India) was used as the source of Ce^{3+}. Aluminum specimens, obtained from Q-Lab Florida, USA, with dimensions 40 mm × 30 mm × 1 mm, were used as anodes after thorough degreasing with acetone. The chemical composition of the aluminum substrate in wt% was Al-98; Mn-1; Si-0.6; Zn-0.1; Cu-0.05–0.3. The sol was synthesized in two parts and then mixed together to obtain the nanocomposite sol. In the first step, zirconium-n-propoxide was complexed with methacrylic acid in a mole ratio of 1 : 1 under vigorous stirring to reduce the hydrolysis rate of Zr-n-propoxide. In the second step, GPTMS and water mixture was stirred with 0.1 N HCl for prehydrolysis of GPTMS and after ensuring hydrolysis of GPTMS, both parts were mixed and stirred continuously for 5 h. The mole ratio of GPTMS : Zr-n-propoxide was 3 : 1. The sol (henceforth abbreviated as GZ) was then appropriately diluted using 2-butoxy-ethanol and cerium nitrate was added so that final concentration of Ce^{3+} in the sol was 0.01 M.

Anodization/Deposition

The coating was deposited electrochemically using the inorganic-organic hybrid sol (GZ sol), along with Ce^{3+} as the source of self-healing material. A programmable current source (Keithley 224) was used for supplying the current during the deposition process. The process of anodization/deposition was monitored with current source as well as withdrawal speed. The pH of the sol was varied in three steps, namely, 4, 8, and 10. The current density was varied between 1.2 and 4 mA/cm^2 in three steps, namely, 1.2 mA/cm^2, 2.4 mA/cm^2, and 4 mA/cm^2, and the time of immersion of substrates in the sol was also varied for 1, 5, and 10 minutes. The optimum conditions were found to be the deposition of a sol maintained at pH 8–8.5 using a current density 4 mA/cm^2 for a deposition time of 10 minutes. After the layer formation on the substrate, it was removed from the sol tank using a dip coater with 1 mm/s withdrawal speed. The coatings were cured in air at 130°C for 1 h. Some of the coatings were also cured in air at 300°C for 1 h. For the sake of comparison, normal dip coating of the sol was also carried out on some of the substrates using 1 mm/s withdrawal speed and coatings were cured at 130°C for 1 h.

Characterization

The coating thickness, morphology, and composition were analyzed by SEM/EDAX, using a scanning electron microscope (Hitachi model S/3400N) and the crystalline nature was ascertained using an X-ray diffractometer (Bruker D8 AXS Advance X-ray diffractometer), using both normal incidence (NI) and grazing angle incidence (GI) of 1 degree. The adhesion of the coatings was tested according to ASTM

D3359-02. Scratch test is a simple and rapid method to characterize the coatings, but results obtained are influenced by various factors such as coating thickness, mechanical properties of the substrate, interfacial bond strength, and test conditions such as scratch speed, load, and indenter tip radius. Scratch test was carried out using a microscratch tester (Revtest, CSM make). A Rockwell diamond indenter with a tip radius of 200 μm and a progressive loading method from a minimum of 0.9 N at the point of contact to a maximum of 10 N normal load at the end of scratch length were employed for evaluating the coated substrates. Scratch length, loading rate, and scratching speed were fixed at 6 mm, 7.58 N/min, and 5 mm/min, respectively. In the present investigation, the scratch tester tip was brought in contact with the coated surface and the sample was moved at a constant speed, while the tip normal load was progressively increased to the set maximum value. The output was measured in terms of acoustic emission, penetration depth, and tangential frictional force. The scratch tester has a tip, which is placed with a controlled scratch load F_z on the surface to be tested. When the sample is scratched, the tip is stationary and sample moves. The resulting frictional force F_x can be monitored, while the scratch track is generated.

Electrochemical tests were carried out using an electrochemical interface (Solartron SI 1287) with an impedance analyzer (Solartron SI 1260). The corrosion test cell had the classic configuration of three electrodes, platinum electrode as counter electrode, a saturated calomel electrode as a reference electrode, and the uncoated/coated aluminum substrate as the working electrode. Polarization studies of the coated/uncoated substrates were carried out at 25°C in a N_2 purged 3.5% NaCl solution with an exposure time of 1 h and 24 h. Potentiodynamic scans were recorded by applying potentials from −1.6 V to 0.0 V with a scan rate of 1 mV/s. The electrochemical impedance scan was carried out using an AC signal of 10 mV amplitude applied over a bandwidth from 100 kHz to 0.03 Hz. The corrosion behavior was also tested by neutral salt spray tests, using 5% NaCl solution according to ASTM B117.

RESULTS AND DISCUSSION

XRD Analysis

Structural properties were investigated with both normal incidence XRD and GIXRD, the latter being a surface sensitive technique. The grazing incidence method was employed due to the large penetration depth of X-rays into the bulk during normal incidence, which tends to add diffraction peaks from the underlying crystalline matrix of the substrate to that of the coating on the surface, thereby masking the information due to the coating. Figure 1 presents the comparison of the normal incidence XRD patterns of bare aluminum (Figure1 (a)) and that of the coated aluminum (Figure 1(b)), which was cured at 130°C. However, it can be clearly seen that both patterns look similar. Due to high penetration depth of X-rays, the information from the coating has been masked due to the highly crystalline nature of the metallic substrate. Figure 1(c) shows the XRD pattern obtained from the GIXRD analysis. Here, it can be clearly seen that there is an initial hump, which is an indication of presence of an amorphous phase along with sharp peaks that indicate presence of a crystalline material. Since the peaks do not coincide with those of the aluminum substrate (JCPDS 04-0787), it can be inferred that these peaks are due to a different phase in the coating, which correspond to an aluminosilicate with the formula Al_2SiO_5 (JCPDS 44-0027) The XRD analysis shows that the formed material during the in situ anodization/sol-gel deposition process is crystalline on an as-deposited sample, even at room temperature. The amorphous phase in the coating could be silica and/or zirconia from the GZ sol.

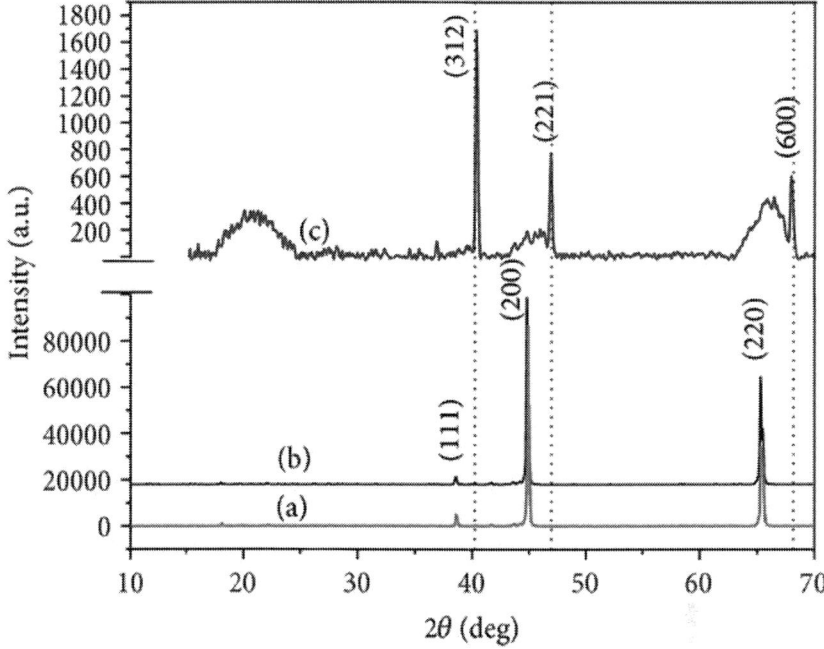

Figure 1: Comparison of XRD patterns of (a) bare aluminum, (b) coated aluminum cured at 130°C, both acquired at normal incidence and (c) that of coated aluminum, cured at 130°C, acquired at GI of 1 degree.

SEM/EDAX Analysis

An image of cross section of the coated aluminum substrate is shown in Figure 2(a). The coating is seen to be 15–20 microns thick and the EDAX spectrum of the coating is shown in Figure 2(b). It can be seen from Figure2 (a) that the coating is highly dense in nature. The SEM images of the surface of the coated sample as shown in Figure 3 confirm that the coating is dense with negligible surface porosity. The coatings cured at 130°C (as shown in Figure 3(a)) do not show any defects. However, when the coatings are heat-treated at 300°C, the magnified image of the surface (inset shown in Figure 3(b)) confirms that cracks had appeared in the coating. Hence, it was concluded that a low temperature of 130°C for curing was sufficient to remove the residual solvent, densify, and generate a defect-free coating.

(a)

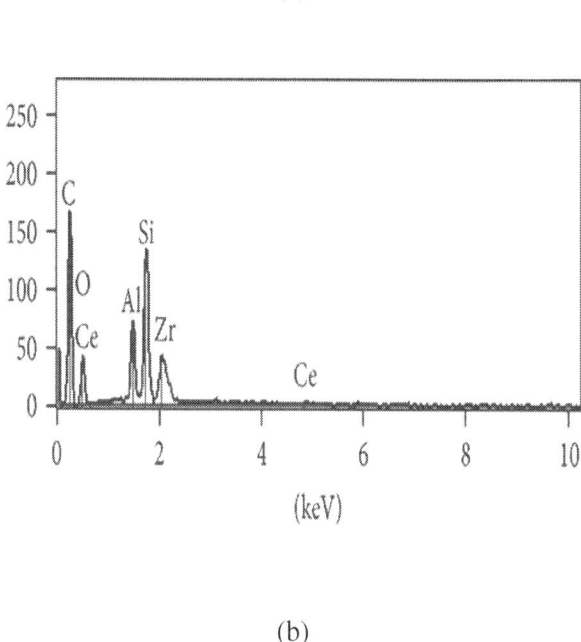

(b)

Figure 2: (a) SEM image of cross section of coated Al. (b) EDAX spectrum of the indicated area of (a).

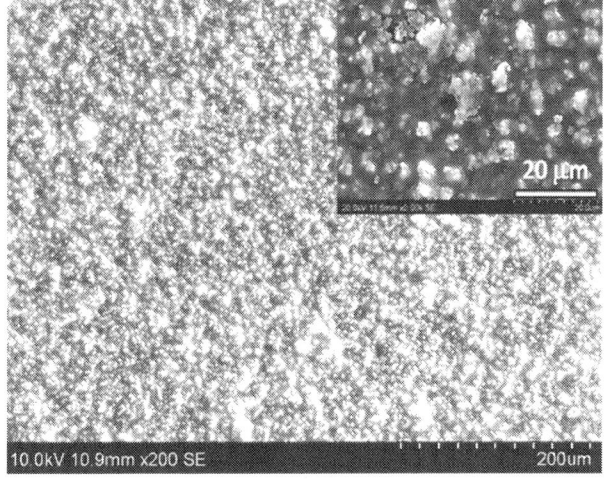

(a)

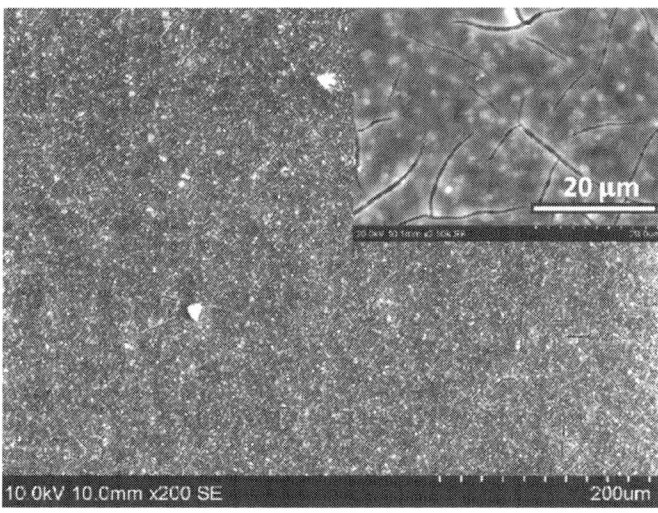

(b)

Figure 3: SEM images of the surface of coated substrate (a) cured at 130°C and (b) cured at 300°C.

Scratch and Adhesion Test

The onset load for coating cracking is commonly referred to as the critical load and is associated with the rise in values of coefficient of friction and acoustic emission. In the present case, the frictional load and the acoustic emission were not clearly distinguishable as the coatings started to crack at about 2 N normal loads. Hence, morphology of the scratch was considered to rank the coatings for simple dip coated and electrolytically deposited coatings that were abbreviated as SD and ED, respectively.

From the scratch test results, the simple dip coated specimen started showing visible cracks and delamination even at the bare minimum load of 0.9 N applied during point of contact, whereas there were no cracks in the electrolytically coated specimens till the normal load increased to 2 N. The loads mentioned here are the average values of three scratch tests carried on each sample at different locations. These are the critical loads for the respective coating conditions. Normally, small fractures are generated across the scratch length with increasing load at certain periodicity. However, the coatings still stay adhered to the substrate till a second critical load. But when the normal load rises further, coating delaminates forming debris on the sides of the scratch. Scratch tracks are visible even to the naked eye due to the scattering of light by the defects along the width and/or depth of the scratch. While the SD sample has failed in classic brittle fracture as can be discerned from Figure 4(a), ED failed according to classic ductile fracture as seen in Figure 4(b). The results from the scratch tests can be further understood by examining the morphology of the ED samples by SEM analysis. The coatings have a microlevel roughness with extensive porosity due to dendrite-like structure as seen from the inset of Figure 3(a). But these dendrites and the pores are covered with a layer of the silica-zirconia formed from the sol. These dendrites have further submicron and nanoparticle structure resulting from the ED process. Such structure acts like a reinforcement and provides necessary improved tensile strength. Hence, the ED coatings are able to resist the brittle fracture at lower load unlike a normal sol-gel coating, though the coating material is the same. Accordingly, the dendrite structure in case of the former is rougher due to enhanced height of such features. The adhesive properties were tested for ED coatings and were found as 5B (0% removal after adhesion test), which implies that the coating has excellent adhesion to the substrate.

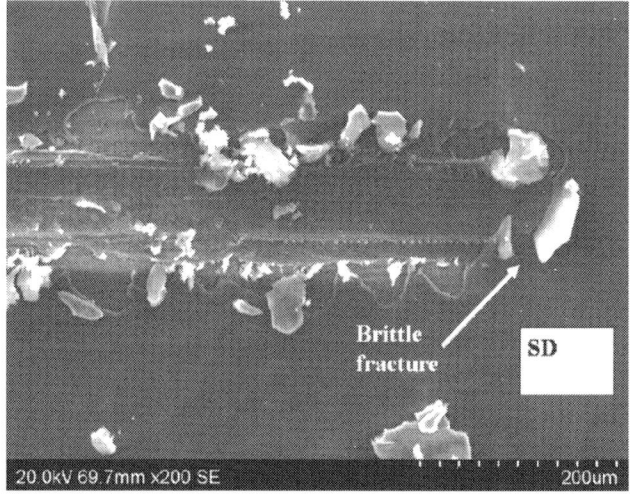

(a)

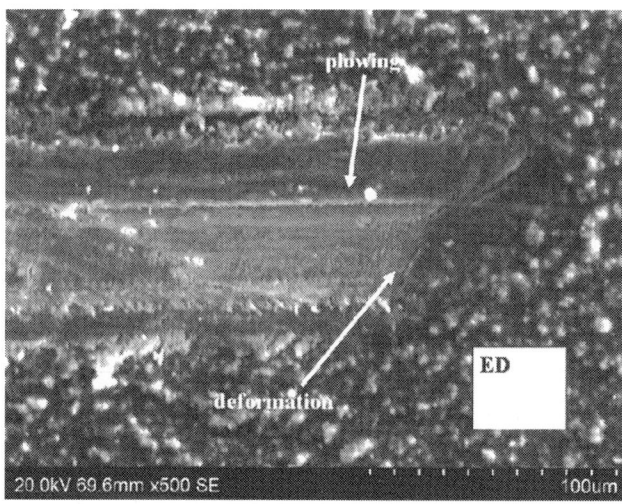

(b)

Figure 4: SEM images of scratch (a) simple dip coated (SD) and (b) electrolytically deposited (ED) aluminum.

Corrosion Tests

Neutral Salt Spray Tests (NSST)

The photographs of coated and bare substrates after NSST are shown in Figure 5. The substrate panels were scratched prior to exposure to salt spray to accelerate the corrosion process. The bare aluminum panel totally failed in 48 hours, while the coating in the normal dip coated substrate started to peel off after 48 hours and the scratched area was filled with the corrosion product. In case of electrochemically anodized cum deposited substrate (ED), no corrosion was found to occur even after 400 hours. The performance of the ED coating as a barrier coating is far superior to simple dip coated aluminum and is capable of rendering a long-term corrosion protection to the aluminum substrate and, hence, further corrosion resistance property measurements like potentiodynamic polarization and electrochemical impedance studies were carried out only on ED samples and not on the SD samples.

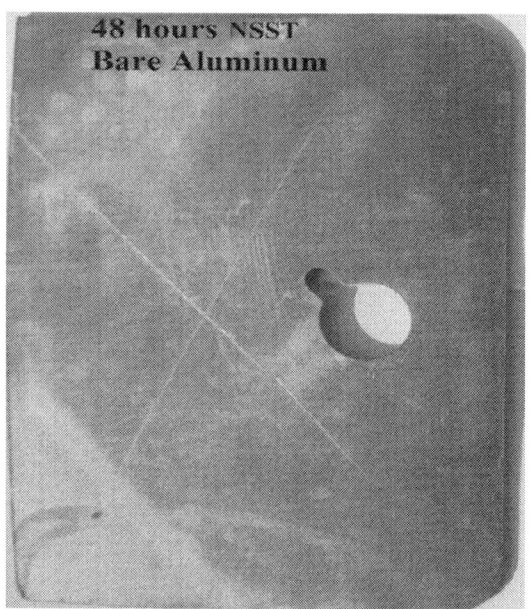

(a)

(b)

(c)

Figure 5: Photographs of samples after salt spray test: (a) after 48 hours—bare aluminum substrate, (b) after 48 hours—dip coated substrate with 3 mm/s withdrawal speed, and (c) after 400 hours—electrolytically deposited substrate.

Potentiodynamic Polarization Measurements

The data obtained from the potentiodynamic polarization studies carried out after 1 h and 24 h exposure to 3.5% NaCl, on the electrolytically coated aluminum and bare aluminum substrates are presented in Figure 6and the results obtained after R_p fitting the data are presented in Table 1. It could be seen that the electrolytically coated Al provided a good barrier protection to the aluminum substrate, since the corrosion current density, i_{corr}, was lower when compared to that for the bare aluminum substrate. The corrosion potentials were lower for the coated aluminum substrates when compared to those for the bare Al substrate. The reduction in the open circuit potential (OCP) could be attributed to the effective suppression of the cathodic reaction due to the reason that SiO_2 having a low isoelectric point (IEP = 1.7 − 3.5) leads to a negative surface charge at pH > 2. Since the coatings are exposed to the electrolyte solution 3.5% NaCl that has a pH from 6.5 to 7, which is higher than the IEP, the open circuit potential of the coated substrate is more negative than that of the bare substrate. Though the coatings had lower corrosion potentials than bare Al, they provided a good barrier effect and reduced the corrosion currents.

Table 1: R_p fit data for the results obtained from potentiodynamic polarization measurements

Sample	Ecorr(Volts)	Icorr(Amp/cm2)×10-7	Rp(Ohms/cm2)
Bare Al-1 h	−0.806	64.104	4069.4
Bare Al-24 h	−1.059	15.849	16460
ED-1 h	−1.409	5.247	49714
ED-24 h	−1.377	2.513	1.038E5

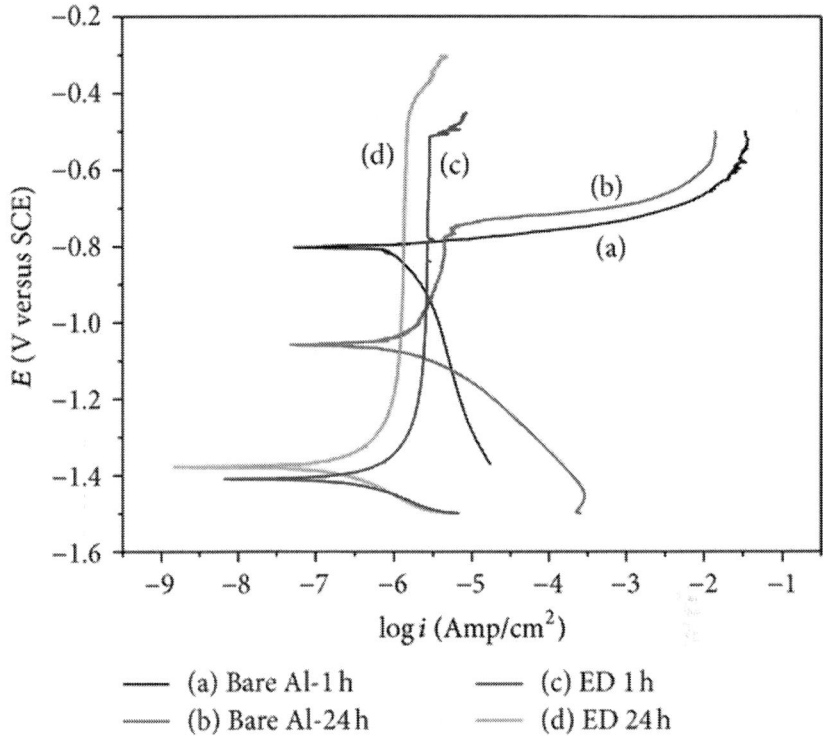

Figure 6: Potentiodynamic polarization plots of bare and coated Al (with ED) after 1 h and 24 h exposure to 3.5% NaCl solution.

In case of only the bare Al, after 1 h exposure, the i_{corr} value is higher in magnitude when compared to that of the 24 h exposure, which implies the formation of an oxide layer during longer exposure time. The oxide layer formation is evident by the passive region formation from −1.0 V to −0.70 V (300 mV), which decreases the rate of corrosion during longer periods of exposure to the electrolyte solution as shown in Figure 6.

It has been reported that the presence of Ce^{3+} in an electrolyte solution reduces the rate of oxygen reduction reaction and E_{corr} shifts to more negative values when compared to that without Ce^{3+} ions [23, 24]. In the present case, the Ce^{3+} ions were present in coating itself. In case of the coated substrates exposed to the electrolyte solution for 1 and 24 h, namely, ED-1 h and ED-24 h, the cathodic arm was shifted to more negative potentials and also lower current densities, when compared to the bare aluminum substrate, indicating a reduction in the rate of oxygen reduction reaction as shown in Figure 6. The corrosion resistance is much higher for the ED samples when compared to that for bare Al for both exposure times. A well-established passive region can be observed for ED samples for 1 h exposure itself and it is still improved for 24 h exposure. This shows that the coatings are highly protective. The 900 mV, that is, the difference between E_{corr} (corrosion potential) and E_c (breakdown potential), is associated with the increased corrosion resistance of passive coatings as shown in Figure 6.

Electrochemical Impedance Spectroscopy (EIS)

EIS is a nondestructive testing technique for evaluation of the barrier properties of the coatings and adhesion to the substrate. The obtained EIS data for the coated substrates are shown as Nyquist plots in Figure 7(a) and were fitted using the equivalent circuit given by two time constants as shown in Figures 7(b) and 7(c).

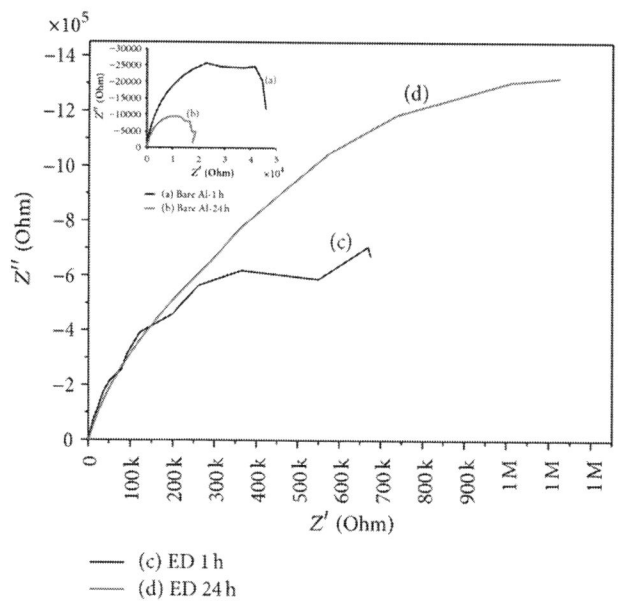

(a)

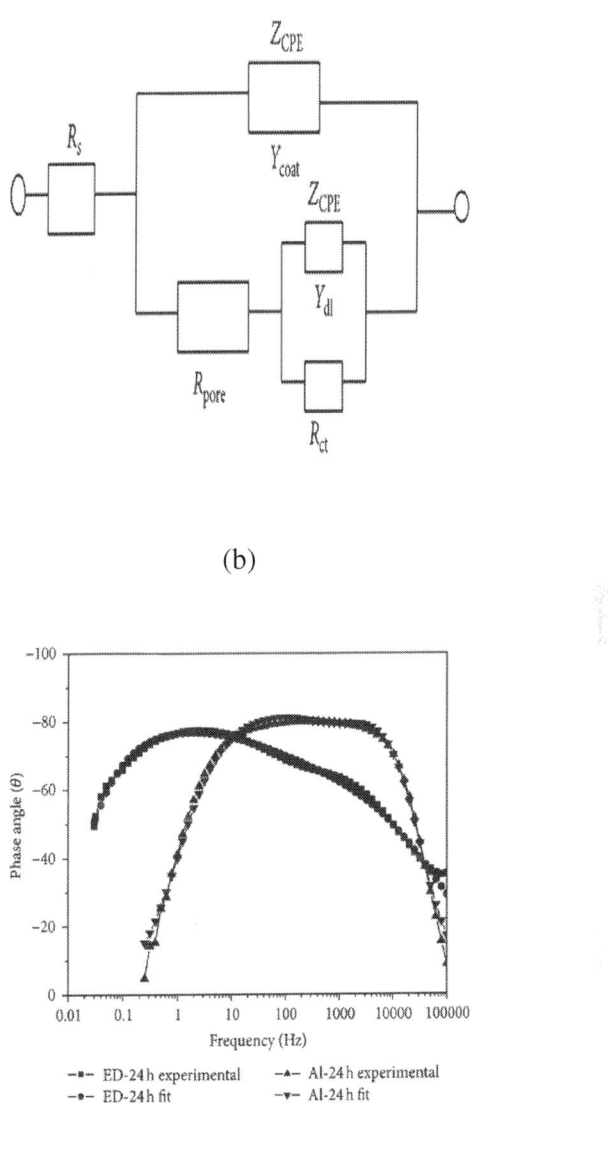

(b)

(c)

Figure 7: (a) Nyquist plots for the coated and bare Al for 1 h and 24 h exposure to 3.5% NaCl, (b) equivalent circuit used for fitting the EIS data of coated substrate, and (c) Bode plot for bare aluminum and ED coating on aluminum after 24 h exposure to 3.5% NaCl.

A constant phase element (CPE) was used instead of an "ideal capacitor" to explain the deviations from ideal behavior. The possible reasons for a nonideal behavior could be due to surface roughness, inhomogeneous reaction rates on the surface, varying thickness, or composition of a coating [25–27]. The time constant at high frequencies is related to the properties of coating and that at medium and lower frequencies is related to the properties of substrate. The impedance of a CPE, that is, Z_{CPE}, could be defined as $Z_{CPE} = 1/Y(j\omega)^n$, where ω is the angular frequency in $rad\,s^{-1}$, Y is the pseudocapacitance, and n is called the CPE exponent, which is associated with the system inhomogeneity. When $n = 1$, the system behaves like a pure capacitor and $Y = C$. When CPE is used to fit the experimental data, $n < 1$; R_s is the resistance of the electrolyte; R_{pore} is the resistance to charge transfer through pores; Y_{coat} is the pseudo-capacitance of coating; Y_{dl} is pseudocapacitance associated with the double layer formed at the metal-electrolyte interface in parallel with charge transfer resistance R_{ct} describing the corrosion of the metal substrate. Table 2 presents the EIS results after circuit fitting of bare and ED samples exposed to 3.5% NaCl for 1 h and 24 h. It can be seen from Figure 7(a) and Table 2 that the R_{ct} values of the ED samples are higher by two orders of magnitude than those of bare samples after 1 h and 24 h exposure which shows that the coating does not allow the electrolyte topenetrate and reachthe substrate. Even the R_{pore} values were also found to be higher (refer to Table 2) for coated samples at both exposed timings than those for bare Al substrate, showing less porosity in the coatings. The higher R_{pore} value of ED sample exposed for 24 h is due to closing of the existing pores by Ce3+ ions which are converted into Ce(OH)3 and Ce(OH)4 precipitates.

Table 2: Results after fitting the EIS data for the bare and ED aluminum substrates after 1 h and 24 h exposure to 3.5% NaCl

Sample	Rs ($\Omega\cdot cm2$)	Ycoat ($S\,s\,n\,cm{-2}$)	n	Ypore ($\Omega\cdot cm2$)	Ydl ($S\,s\,n\,cm{-2}$)	n	Rct ($\Omega\cdot cm2$)	X2
Bare Al-1 h	2.37	7.88E-7	0.97	70.44	4.76E-6	0.83	6.04E4	0.003
Bare Al-24 h	1.54	3.38E-6	0.96	124.7	5.89E-6	0.81	2.11E4	0.011
ED-Al-1 h	27.47	5.56E-7	0.79	163.5	3.20E-6	0.91	2.02E6	0.015

ED-Al-24 h	27.34	1.26E-6	0.83	637.2	8.07E-7	0.86	4.34E6	0.011

CONCLUSIONS

The novel process of in situ anodizing/sol-gel coating deposition substantially improved the corrosion resistance properties of the aluminum substrate and as expected, this one-step anodizing/ deposition process circumvents the deterioration of mechanical properties of substrate due to the process being carried out at room temperature process which does not need high temperature curing. The SEM studies suggest that a low temperature of 130°C is sufficient for curing to obtain crack-free coatings. Since this coating is highly stable in salt spray test, it has considerable potential as future alternatives for nontoxic chromate-free conversion coatings.

ACKNOWLEDGMENTS

The authors gratefully acknowledge the constant support and encouragement provided by Dr. G. Sundararajan and Dr. G. Padmanabham, ARCI, Hyderabad, India. The authors also would like to acknowledge Mr. G. V. R. Reddy for the SEM analysis and Mr. A. Ramesh for the technical support.

REFERENCES

1. M. P. Ryan, D. E. Williams, R. J. Chater, B. M. Hutton, and D. S. McPhail, "Why stainless steel corrodes," Nature, vol. 415, no. 6873, pp. 770–774, 2002. · ·

2. I. Betova, M. Bojinov, T. Laitinen, K. Mäkelä, P. Pohjanne, and T. Saario, "The transpassive dissolution mechanism of highly alloyed stainless steels: I. Experimental results and modelling procedure,"Corrosion Science, vol. 44, no. 12, pp. 2675–2697, 2002. · ·

3. S. A. Kulinich and A. S. Akhtar, "On conversion coating treatments to replace chromating for Al alloys: recent developments and

possible future directions," Russian Journal of Non-Ferrous Metals, vol. 53, no. 2, pp. 176–203, 2012. · ·

4.   V. Moutarlier, M. P. Gigandet, and J. Pagetti, "Characterisation of pitting corrosion in sealed anodic films formed in sulphuric, sulphuric/molybdate and chromic media," Applied Surface Science, vol. 206, no. 1–4, pp. 237–249, 2003. · ·

5.   S. J. Garcia-Vergara, P. Skeldon, G. E. Thompson, and H. Habazaki, "A tracer investigation of chromic acid anodizing of aluminium," Surface and Interface Analysis, vol. 39, no. 11, pp. 860–864, 2007. · ·

6.   L. E. Fratila-Apachitei, F. D. Tichelaar, G. E. Thompson et al., "A transmission electron microscopy study of hard anodic oxide layers on AlSi(Cu) alloys," Electrochimica Acta, vol. 49, no. 19, pp. 3169–3177, 2004. · ·

7.   X. Liu, J. Gen, A. Yu, G. Wang, L. Song, and X. Zhang, "A study on a submerged membrane bioreactor for municipal wastewater treatment," Advanced Materials Research, vol. 531, pp. 415–418, 2012. ·

8.   A. Jagminas, D. Bigeliend, I. Mikulskas, and R. Tomasiunas, "Growth peculiarities of aluminum anodic oxide at high voltages in diluted phosphoric acid," Journal of Crystal Growth, vol. 233, no. 3, pp. 591–598, 2001. ·

9.   R. Narayanan and S. K. Seshadri, "Phosphoric acid anodization of Ti–6Al–4V—structural and corrosion aspects," Corrosion Science, vol. 49, no. 2, pp. 542–558, 2007. ·

10.   S. J. Garcia-Vergara, P. Skeldon, G. E. Thompson, and H. Habazaki, "Pore development during anodizing of Al-3.5 at.%W alloy in phosphoric acid," Surface and Coatings Technology, vol. 201, no. 24, pp. 9506–9511, 2007. · ·

11.   F. L. Coz, L. Arurault, S. Fontorbes, V. Vilar, L. Datas, and P. Winterton, "Chemical composition and structural changes of porous templates obtained by anodising aluminium in phosphoric acid electrolyte," Surface and Interface Analysis, vol. 42, no. 4, pp. 227–233, 2010. · ·

12.   M. A. Song-jiang, L. Peng, Z. Hai-hui, F. U. Chao-Peng, and K. Ya-fei, "Preparation of anodic films on 2024 aluminum alloy in boric acid-containing mixed electrolyte," Transactions of Nonferrous Metals Society of China, vol. 18, no. 4, pp. 825–830, 2008. ·

13. H.-H. Shih and S.-L. Tzou, "Study of anodic oxidation of aluminum in mixed acid using a pulsed current," Surface and Coatings Technology, vol. 124, no. 2-3, pp. 278–285, 2000. ·

14. M. A. Páez, J. H. Zagal, O. Bustos, M. J. Aguirre, P. Skeldon, and G. E. Thompson, "Effect of benzotriazole on the efficiency of anodizing of Al-Cu alloys," Electrochimica Acta, vol. 42, no. 23-24, pp. 3453–3459, 1997.

15. T. Takenaka, H. Habazaki, and H. Konno, "Formation of black anodic films on aluminum in acid electrolytes containing titanium complex anion," Surface and Coatings Technology, vol. 169-170, pp. 155–159, 2003. · ·

16. A. Da Forno, M. Bestetti, N. Lecis, S. P. Trasatti, and M. Trueba, "Anodic oxidation and silane treatment for corrosion protection of AM60B magnesium alloy," Materials Science Forum, vol. 690, pp. 413–416, 2011. · ·

17. F. Zucchi, V. Grassi, A. Frignani, C. Monticelli, and G. Trabanelli, "Influence of a silane treatment on the corrosion resistance of a WE43 magnesium alloy," Surface and Coatings Technology, vol. 200, no. 12-13, pp. 4136–4143, 2006. · ·

18. K. Watanabe, M. Sakairi, H. Takahashi, K. Takahiro, S. Nagata, and S. Hirai, "Anodizing of aluminum coated with silicon oxide by a sol-gel method," Journal of the Electrochemical Society, vol. 148, no. 11, pp. B473–B481, 2001. · ·

19. X. Du and Y. Xu, "Formation of Al_2O_3-$BaTiO_3$ nanocomposite oxide films on etched aluminum foil by sol-gel coating and anodizing," Journal of Sol-Gel Science and Technology, vol. 45, no. 1, pp. 57–61, 2008. ·

20. J. G. Kaufman, Properties of Aluminum Alloy, ASM International, Materials Park, Ohio, USA, 2008.

21. D. Wang and G. P. Bierwagen, "Sol-gel coatings on metals for corrosion protection," Progress in Organic Coatings, vol. 64, no. 4, pp. 327–338, 2009. · ·

22. M. L. Zheludkevich, J. Tedim, and M. G. S. Ferreira, "'Smart' coatings for active corrosion protection based on multi-functional micro and nanocontainers," Electrochimica Acta, vol. 82, pp. 314–323, 2012. ·

23.   B. R. W. Hinton, "Corrosion inhibition with rare earth metal salts," Journal of Alloys and Compounds, vol. 180, no. 1-2, pp. 15–25, 1992. ·

24.   E. A. Matter, S. Kozhukharov, M. Machkova, and V. Kozhukharov, "Comparison between the inhibition efficiencies of Ce(III) and Ce(IV) ammonium nitrates against corrosion of AA2024 aluminum alloy in solutions of low chloride concentration," Corrosion Science, vol. 62, pp. 22–33, 2012. ·

25.   C. A. Schiller and W. Strunz, "The evaluation of experimental dielectric data of barrier coatings by means of different models," Electrochimica Acta, vol. 46, no. 24-25, pp. 3619–3625, 2001. ·

26.   W. H. Mulder, J. H. Sluyters, T. Pajkossy, and L. Nyikos, "Tafel current at fractal electrodes: connection with admittance spectra," Journal of Electroanalytical Chemistry, vol. 285, no. 1-2, pp. 103–115, 1990. · ·

27.   C.-H. Kim, S.-I. Pyun, and J.-H. Kim, "An investigation of the capacitance dispersion on the fractal carbon electrode with edge and basal orientations," Electrochimica Acta, vol. 48, no. 23, pp. 3455–3463, 2003. · ·

Hydrogenated Microstructure and its Hydrogenation Properties: A Density Functional Theory Study

M. Abdus Salam, Bawadi Abdullah,
and Suriati Sufian

Chemical Engineering Department, Universiti Teknologi PETRONAS,
Bandar Seri Iskandar, 31750 Tronoh, Perak Darul Ridzuan, Malaysia

ABSTRACT

The relationship between microstructure and hydrogenation properties of the mixed metals has been investigated via different spectroscopic techniques and the density functional theory (DFT). FESEM and TEM analyses demonstrated the nano-grains of Mg_2NiH_4 and MgH_2 on the hydrogenated microstructure of the adsorbents that were confirmed by using XPS analysis technique. SAED pattern of hydrogenated metals attributed the polycrystalline nature of mixed metals and ensured the hydrogenation to Mg_2NiH_4 and MgH_2 compounds. Flower-like rough surface of mixed metals showed high hydrogenation capacity. The density functional theory (DFT) predicted hydrogenation properties; enthalpy and entropy changes of hydrogenated microstructure of MgH_2 and Mg_2NiH_4 are -62.90 kJ/mol, -158 J/mol·K and -52.78 kJ/mol, -166 J/mol·K, respectively. The investigation corresponds to the hydrogen adsorption feasibility, reversible range hydrogenation thermodynamics, and hydrogen desorption energy of 54.72 kJ/mol. DFT predicted IR band for MgH_2 and Mg_2NiH_4 attributed hydrogen saturation on metal surfaces.

INTRODUCTION

The global energy demand is increasing in parallel with the population growth, economic expansion, and increasing demand of mobility. Hydrogen is known as an abundant source of clean energy carrier. There are few ways to utilize the hydrogen as source of energy. Metal-hydrogen system demonstrated the best option for hydrogenated system due to the higher hydrogen density in the system than liquid or gas phases [1]. Higher hydrogen density, reversible thermodynamics, low desorption temperature, and low production cost are vital requirements for metal-hydrogen system. Magnesium based mixed oxides are potential candidate to meet the above requirements [2]. The drawbacks of hydrogenation in magnesium based alloys are [3] stability of the MgH_2 phase, formation of a surface oxide(s), slow dissociation of hydrogen at the metal surface, and slow diffusion of hydrogen through MgH_2 grain. The problems can be overcome by developing microstructure that reduces the diffusion path and creates enough active site for dissociation [2, 3]. Grain boundaries provide active nucleation sites in hydride formation and decomposition of the hydride phases. Reduction of grain size or crystalline size of the materials enhances the hydrogen diffusion that moves faster along the grain boundaries.

Different strategies have been investigated to achieve reversible range hydrogen adsorption thermodynamics and kinetics. Homogenous and ultrafine microstructure of adsorbent is one of the important techniques to enhance the hydrogenation capacity and to achieve favorable thermodynamics [4]. Second technique is considered as addition of catalytic elements or rear earth metals or transitional metals to adsorbent. Transition metal oxides are an important catalyst in Mg-based adsorbent to get high hydrogenation capacity [5]. Mg_2Ni alloy demonstrates higher hydrogenation kinetics and desorption at lower temperature because of the effect of transitional metal such as nickel [6]. Coprecipitation under low supersaturation synthesis method produces homogenously metals dispersed hydrotalcite [7]. The materials show high catalytic activity for hydrogenation and flexible to form different microstructure by optimizing the preparation conditions and metals.

Many DFT based studies have been reported for magnesium based hydrogen storage adsorbents [8, 9]. The density functional theory (DFT) based QM (quantum mechanical) descriptors describe adsorption properties such as adsorption thermodynamics and kinetics, interaction energy, and adsorption and desorption energy that support to observe the hydrogenation mechanism and properties [10] of the materials. Noncovalent week interaction energies can be determined precisely using M05-2X functional [11].

Microstructure influences hydrogenation or dehydrogenation properties strongly [12]. It is a novel initiative to make a clarification of hydrogenation on reduced mixed metal oxides microstructure via experimental and theoretical way. In this study, hydrogenated microstructure is investigated through electron microscopic techniques and studied hydrogenation/dehydrogenation properties using the density functional theory (DFT) to establish a relation between hydrogenation and microstructure of the material.

EXPERIMENTAL SECTION

Experimentation

Hydrotalcite based mixed oxides containing Mg, Ni, and Al were prepared by using the conventional coprecipitation method under low supersaturation method as described elsewhere [13]. The synthesis pH of 10 was optimized to get the flower-like rough surface. Optimized synthesis temperature of 55°C and molar ratio of metals were used as Mg : Ni : Al = 2 : 1 : 1. The fresh developed materials were calcined at 650°C for two hours to convert the mixed metals oxides.

The hydrogen adsorption equilibrium of mixed oxide was measured at the conditions of temperature (30°C) and pressure (0–760 mmHg) using a micrometrics ASAP 2020C sorptometer. A 20 mg sample was degassed using both preparation port and sample port. It was reduced by H_2 gas purging at 20 mL/min, with a temperature ramping of 10°C/min up to 850°C, which further cool down to 30°C under vacuum condition. 99.99% hydrogen gas was used to inlet the adsorption unit. Hydrogen adsorption isotherm analyses were carried out at 30°C with a pressure of 0–760 mmHg. Reversible adsorption isotherms were collected by performing a 30-minute evacuation at the experiment temperature following the total adsorption isotherm.

The microstructure of hydrogenated mixed metals was analyzed using field emission scanning electron microscopy (FESEM) technique. CARL Zeiss Supra 55VP FESEM instrument equipped with the Oxford INCA 400 EDX microanalysis system with an operating voltage in the range of 0.1–30 kV was used. The system was vacuumed before analyses and the mixed metals were introduced in the sample grid.

The microstructure of hydrogenated adsorbent was investigated using a CARL Zeiss LIBRA 200FE, TEM using ultrasonically dispersed samples in isopropanol. Selected area electron diffraction (SAED) patterns were collected for hydrogenated single sheet. The diffraction rings of SAED pattern were indexed using the diameter of the rings and calculating the ratio of those diameters (r_n/r_1) for different phases.

The surface composition of the hydrogenated adsorbent was analysed by X-ray photoelectron spectroscopy (XPS) using a Perkin Elmer 1257 model, operating at a base pressure of 3.5×10^{-6} Pa at 300 K with a nonmonochromatized Al-Kα line at 1486.6 eV and a hemispherical sector analyzer.

Computational Methods

The hydrogenated clusters or hydrides were optimized (geometrically) using M05-2X functional [11] along with the 6-311+G(d,p) basis set [14]. All computations were performed using Gaussian 09W computational program package [15]. Frequency calculations were conducted on optimized geometries of clusters to ensure the minima on potential energy surface (PES). Cluster descriptors such as chemical potential [16], strength of charge transfer [17], and electrophilicity [18] were calculated by using finite approximation. Energy of interaction (ΔE_i), hydrogenation energy (ΔE_{Hyd}), and average desorption energy (ΔE_{DE}) [19] per hydrogen molecule were calculated using different equation as stated by Salam et al. [10]. Thermodynamics parameters such as heat of reaction and free energies have been calculated using computed energies of reactants and products [20]; The IR spectrum of hydride clusters was generated using computational software (Gaussian 09W) via frequency calculation.

RESULTS AND DISCUSSION

Experimental Analyses

Hydrogen Adsorption Mechanism

Isothermal adsorption analysis predicted the hydrogen storage capacity of 3.9 wt% hydrogen of the reduced mixed metals oxides containing magnesium, nickel, and aluminium [13]. A hydrogenation mechanism and causes of hydrogenation can be explained by pointing out different factors.

The surface of s-orbital electron and d-orbital electrons of atom plays a key role in the hydrogen dissociation on the metals surface. H_2 molecules come to the metals surface in the hydrogenation process. The molecular orbitals of hydrogen ($1s^2$) start to overlap on the metals surface s-orbital electrons. According to the Pauli Exclusion Principle, hydrogen s-orbital electron repels due to the fully occupied valence electron of metals (Mg: $1s^2 2s^2 2p^6 3s^2$), resulting in the energy barrier. The metals with d-orbital electrons contribute in the dissociation of H_2. H_2 comes closer to the metal surface; the charge transfers from the H_2 s-orbital to the d-orbital of metal (Ni (28): $1s^2 2s^2 2p^6 3s^2 3p^6 3d^8 4s^2$) surface and simultaneously a back-donation of charge from the surface d-orbital to the H_2 antibonding state. Thus, the interaction between d-orbital of nickel and s-orbital of H_2 enhances the H_2 dissociation on the adsorbent surface with a small energy barrier. The main group element of aluminum (Al (13): $1s^2 2s^2 2p^6 3s^2 3p^1$) valence electron is three (3).

The electron affinity of Al is −44 kJ/mol and shows low affinity to H. Since the p-state of Al can accommodate electron more, it competes with H (hydrogen) for the valence electrons of Mg (Mg (12): $1s^2 2s^2 2p^6 3s^2$). The electron affinities of magnesium and hydrogen are 21 kJ/mol and −74 kJ/mol. Therefore, hydrogen binds with weak interaction (lower bonding strength) with the Mg or Mg_2Ni and forms reversible hydride of Mg_2NiH_4 and MgH_2. Aluminum also works as cofactor to heat transfer quickly for better decomposition of hydrides from microstructure.

Microstructure of Mixed Metals Oxides

Figure 1 shows the flower-like rough microstructure of the mixed metals oxides. At pH = 10, cations Al^{3+}, Ni^{2+}, and Mg^{2+} precipitated simultaneously as $Mg(OH)_2$, $Ni(OH)_2$, and $Al(OH)_3$ that immediately converted to the $Mg(Ni)Al-CO_3$ structure. The flower-like particle of material can be considered as composed of hexagonal nanosheets which are dependent on isoelectric point of metal hydroxides. The nanosheet or nanoflakes were observed via high resolution FESEM image (not shown here). The nanosheets seem to be flexible and can be wrinkled during precipitation at experimental conditions that help to form flower-like particle. If pH = IEP (isoelectric point), the MgAl-CO_3 nanoparticle is in net zero point charge and the formation of hydrotalcite becomes slow along the 001 plane. In the initial stage, the interfacial energy between the nuclei and the support is low and metal hydroxide nuclei are grown within their 001 plane. Since the reaction is continuous (during synthesis), the subsequent nucleation is grown preferentially on the surface of the previously grown particles. Thus, the growth of the primary nanoparticles turns to flower-like morphology. After calcination, the materials retained its shape. The reason is that the trivalent cations of Al^{3+} substitute the divalent cations of Mg^{2+}, Ni^{2+} to the brucite sheets and converted to periclase-like Mg-Al-O or Mg(Ni)Al-O solid solution. Thus, the brucite-like shape of the crystal remained in the resulting Mg/Ni-Al-O structure. Different synthesis conditions demonstrate different microstructure for this metals combination.

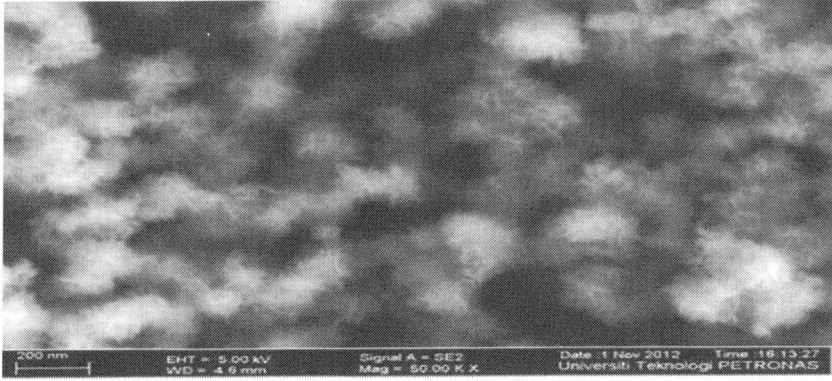

Figure 1: Microstructure of synthesized mixed oxides.

EDX analysis demonstrated the elemental composition of microstructure of the mixed metals oxides in Figure2. Since the material is in mixed oxides form, it shows high amount of oxygen (30 wt%). Magnesium, nickel, and aluminium maintain the ratio (based on wt%) of $2:1:1$.

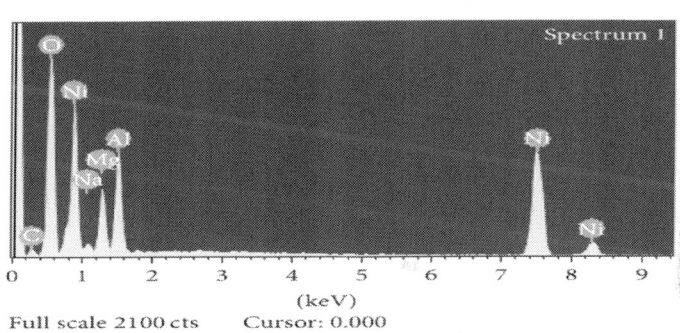

Figure 2: Elemental composition of the material microstructure.

Microstructure of Hydrogenated Mixed Metals

Microstructure of the hydrogenated mixed metals is shown in Figure 3. Hydrogen makes bond with every flake of the flower-like particle, thus forming very small grains. Therefore, well-defined grains attribute the faster hydrogenation and high hydrogen capacity of Mg_2NiH_4 [6].

The size range of hydrogenated grains is in 20–40 nm and synthesized adsorbent form 100–180 nm sizes of flower-like particles in optimized experimental conditions. The rough surfaces of flower-like particles offer active sites to attach hydrogen to each flake and enhance the hydrogenation capacity. The hydride nucleation takes place at the end of flakes including grain boundaries. The hydrogenation rate and hydrogen capacity of material increase rapidly with increasing active site and for the homogenous distribution of catalytic metals of nickel on microstructure. The surface with hexagonal particle demonstrates less hydrogenation in same conditions [13] which is due to the shortage of active site of materials. The -MgH$_2$ is formed at ambient temperature as a rutile structure as observed by Zachariasen et al., [21].

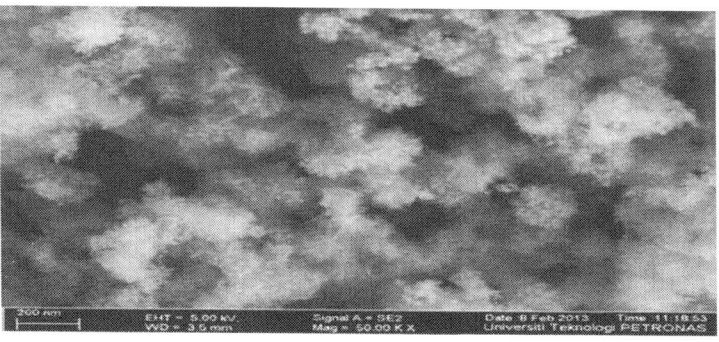

Figure 3: Microstructure of hydrogenated complex.

The TEM image of the adsorbent (Figure 4) shows the flower-like particle which is also observed by FESEM image in Figure 1(a). The newly formed irregular shape of grain adjacent to the flakes of the particle is due to the formation of different phases of hydride at ambient conditions. The adsorbent shows the well-shaped diffusion path and grain boundaries. SAED (selected area electron diffraction) pattern of hydrogenated metals is shown in Figure 5 which attributes the crystal planes of (110), (202), (220), and (330) and corresponding the presence of MgH$_2$, Mg$_2$NiH$_4$, and MgO in crystal. Similar planes of SAED pattern were observed by Zou et al., [22]. Different metals' combination creates a different microstructure due to the structural change and balance band of the adsorbent that influence the hydrogenation on material surface.

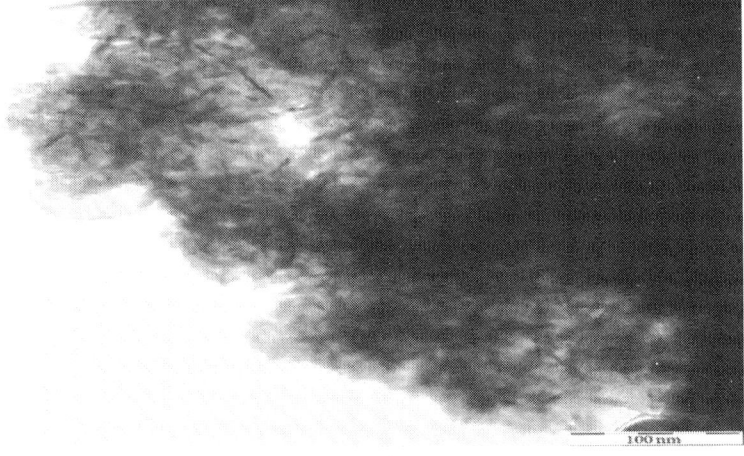

Figure 4: TEM of hydrogenated complex.

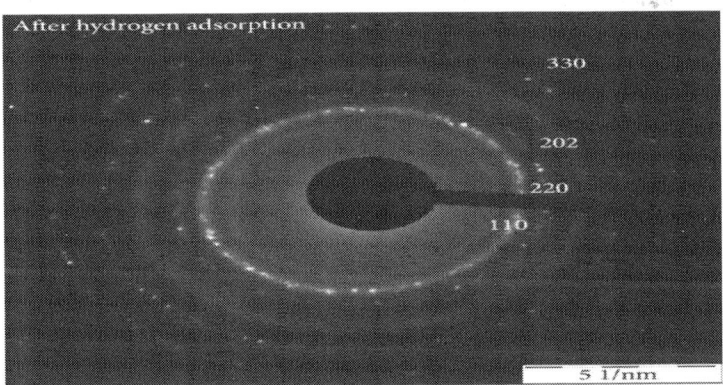

Figure 5: SAED pattern of hydrogenated complex.

Hydrogenated Phases of Adsorbent

Hydrogenated microstructure was analyzed by XPS techniques to confirm the hydride phases and for further studies. Binding energy of hydrides is higher than elemental species, because metals loss their electron in forming hydrides [23]. X-ray photoelectron spectra of Mg (1s), Mg (2p), and Ni (2p) are shown in previous investigation [6]. Hydride phases of MgH_2 and Mg_2NiH_4 can be observed by measuring their higher binding energies as compared to corresponding elements. Combination of magnesium and nickel modifies the electronic structure to Mg_2NiH_4 that enhances the hydrogenation. Mg and Ni change the structure that reduces the formation of Mg $(OH)_2$ and increases hydrogenation [24].

Microstructure Analysis Using DFT

The study highlights the hydrogenation properties of hydrogenated microstructure or hydrogenated phases of MgH_2 and Mg_2NiH_4 that were confirmed by experimental analyses in this study. The geometry of clusters has been optimized successfully and shown in Figure 6. To observe the variation of hydrogenation properties, $Mg-(H_2)_2$ is included along with two hydride clusters of MgH_2 and Mg_2NiH_4. Frequency analyses predict their minima on potential energy surface (PES) and free from imaginary frequencies (Figures 7 and 8). The electronic properties of clusters and feasibility of their application are discussed next.

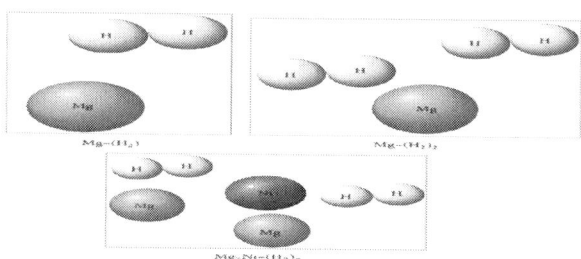

Figure 6: Optimized geometry of $Mg-(H_2)_n$ and Mg_2NiH_4 hydrides.

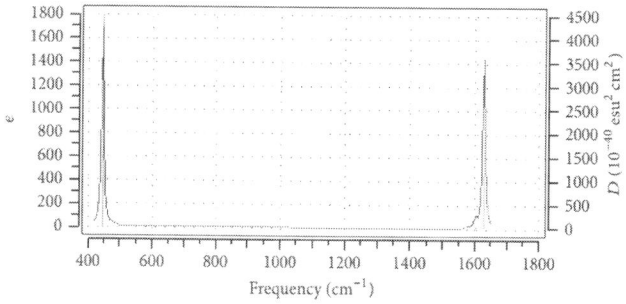

Figure 7: Software generated IR spectroscopy of MgH_2.

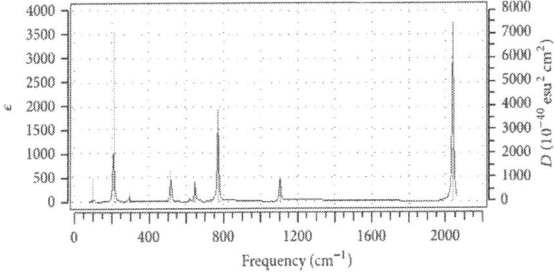

Figure 8: Software generated IR spectroscopy of Mg_2NiH_4.

Total electronic energy and thermodynamical descriptors, hardness, electronegativity, and electrophilicity of hydrides are calculated and shown in Table 1. The increasing trend of electronic energy of Mg-$(H_2)_n$ based on gradual number of captured hydrogen corresponds hydrogenation on metals surface. Chemical hardness determines the clusters strength to donate/accept the charges and the chemical potential represents the electron shifting trend from systems or clusters. Minimum electrophilicity principle (MEP) and maximum hardness principle (MHP) attribute the Mg_2NiH_4 hydride which is more stable than Mg-$(H_2)_n$ [25].

Table 1: Hydride clusters descriptors

Cluster name	E kJ/mol	kJ/mol	kJ/mol	kJ/mol
Mg-(H_2)	-5.280×10^5	86.74	911.92	605.47
Mg-$(H_2)_2$	-5.311×10^5	118.09	376.75	601.00
$Mg_2Ni(H_2)_2$	-5.015×10^6	190.48	383.88	386.83

E: total electronic energy, η: hardness, χ: electronegativity, ω: electrophilicity.

The IR studies on hydrogenated clusters provide two significant properties. IR spectrum of hydrogenated clusters shows that they are free from imaginary frequency indicating that the minima on potential energy surface (PES). Secondly, it confirms the hydrogen saturation on metal surface. IR peaks in the frequency region of 400–2000 cm^{-1} confirm the hydrogen saturation on metal surface of Mg and Ni containing clusters. Linear structure of MgH_2 was produced in the method of UM052X/6-311+G(d,p) calculation and a strong Mg-H antisymmetric stretching frequency mode at 1637 cm^{-1} and a bending frequency mode at 460 cm^{-1} is observed. The inclusion of ground-state Mg (3s^1, S) into hydrogen experiences large energy barrier. Therefore, Mg atom exited to the electronic state of (3s3p, ^{1}P) to overcome the energy barrier and insert into hydrogen. The vibration between 1000 and 1200 cm^{-1} is because of the bridge of Mg-H-Mg bond stretching. Outcome of calculation suggests that stronger interaction for H-Mg-H with hydrogen on metal surface is due to the strong H-H stretching modes induced by MgH$_2$. The band at 1645 cm^{-1} is symmetrical stretching mode of H-Ni-H. Intermetallic ternary hydride Mg$_2$Ni-(H$_2$)$_2$ (=Mg$_2$NiH$_4$) displayed the vibrational band (Figure 8) at 222, 528, 650, 7685, 1100, and 2030 cm^{-1} which includes the common band of Ni-(H$_2$)$_n$; here, n = 1 or 2 [26].

Different energy parameters have been investigated for Mg-(H$_2$)$_n$ and Mg$_2$NiH$_4$ and shown in Table 2. Interaction energies between hydrogen and metal cluster increase with gradual number of captured hydrogen. Hydrogenation and interaction energies are increasing based on gradual addition of hydrogen. These phenomena correspond to the favorable hydrogenation on metals. Table 2 shows that adsorption energy values are higher than 50 kJ/mol and indicate chemisorption range hydrogenation (strongly bonded). According to the calculated data, (Table 2) interaction energy corresponds to the fact that the interaction between metals and hydrogen is electrostatic range. Magnesium shows higher affinity with hydrogen than nickel. Addition of nickel decreases strength of bonding between magnesium and hydrogen. The average desorption energies (Table 2) of the hydrides attribute that desorption may occur at higher condition (temperature, pressure) for the less adsorbed hydrogen and maximum captured hydrogen decompose at low temperature. Predicted hydrogen desoprtion energies (or decomposition energy) of MgH$_2$ and Mg$_2$NiH$_4$ are 79.31 kJ/mol and 54.72 kJ/mol which is one of the important concerns for reversible hydrogenated materials and those are close to the calculated desorption energy value of 76 kJ/mol (for MgH$_2$) by Swart et al., [27]. Reaction electrophilicity values of hydrides are decreasing upon increasing number of hydrogen captured. This phenomenon is very favorable for high hydrogenation [28].

Table 2: Energy descriptors of clusters

Cluster name	E_i kJ/mol	E_{Hyd} kJ/mol	E_{des} kJ/mol	kJ/mol
$Mg–(H_2)$	−76.13	76.13	79.31	212.46
$Mg–(H_2)_2$	−43.93	87.84	78.00	104.43
$Mg_2Ni–(H_2)_2$	−25.11	50.22	54.72	65.95

ΔE_i: interaction energy, ΔE_{Hyd}: adsorption energy, ΔE_{des}: desorption energy, : reaction electrophilicity.

The thermodynamic descriptors of hydride phases such as change of Gibb's free energies (ΔG_r), enthalpy change (ΔH_r), and entropy change (ΔS_r) were computed to ensure the feasibility of hydrogenation on metals surfaces. The negative (−) values of Gibb's free energies (ΔG_r), shown in Table 3, ensure the favorable hydrogenation phenomenon at room temperature and indicate the impetuous behavior of hydrogenation. DFT predicted values of hydrogenation enthalpy and entropy changes of hydrogenated clusters of MgH_2 and Mg_2NiH_4 are −62.90 kJ/mol, −158 J/mol·K and −52.78 kJ/mol, −166 J/mol·K, respectively, that are close to reversible hydrogen storage materials thermodynamics values of −40 kJ/mol and −130 J/mol·K [6]. Negative enthalpy (ΔH_r) of hydrides indicates that the hydrogenation is exothermic. Decreasing trend of entropy corresponds to the limitation of hydrogen movement on surface due to the less degrees of freedom of hydrogen. The reaction constant values are in the order of ($A \times 10^{13}$) that attribute to the slow hydrogenation on this materials surface.

Table 3: Thermodynamics and kinetics descriptors

Cluster name	H_r kJ/mol	G_r kJ/mol	S_r J/mol·K	k S^{-1}
$Mg–(H_2)$	−62.90	−5.64	−158.18	6.0×10^{13}
$Mg–(H_2)_2$	−58.85	−1.42	−159.13	1.1×10^{13}
$Mg_2Ni–(H_2)_2$	−52.78	−13.41	−166	1.3×10^{15}

ΔH_r: reaction enthalpy, ΔG_r: Gibb's free energy, k: rate of adsorption at 303 K.

CONCLUSIONS

The density functional theory (DFT) study has explained the mechanism of the hydrogenation and the hydrogenation properties of microstructure. Variation of the thermodynamics and energy descriptor values upon increasing captured hydrogen molecules correspond to the promising activities of the hydride microstructure and their feasible hydrogenation properties. Experiment and DFT based investigation demonstrated similar IR frequencies, which confirm the hydride formation of MgH_2 and Mg_2NiH_4. Isothermal adsorption analysis predicted high hydrogenation capacity which is higher than the stoichiometric capacity of Mg_2NiH_4 (3.6 wt% hydrogen). Polycrystalline and flower-like interactive rough microstructure is an important material's properties that can be formed in an optimal condition to enhance the hydrogenation.

ACKNOWLEDGMENTS

The authors acknowledge gratefully the financial support of this study by the FRGS Grant 158-200-092, Malaysia, and the computational facility provided by University of Malaya.

REFERENCES

1.   U. Eberle, G. Arnold, and R. von Helmolt, "Hydrogen storage in metal-hydrogen systems and their derivatives," Journal of Power Sources, vol. 154, no. 2, pp. 456–460, 2006. · ·

2.   H. Shao, G. Xin, J. Zheng, X. Li, and E. Akiba, "Nanotechnology in Mg-based materials for hydrogen storage," Nano Energy, vol. 1, no. 4, pp. 590–601, 2012. · ·

3.   I. P. Jain, C. Lal, and A. Jain, "Hydrogen storage in Mg: a most promising material," International Journal of Hydrogen Energy, vol. 35, no. 10, pp. 5133–5144, 2010. · ·

4. W.-J. Song, J.-S. Li, T.-B. Zhang et al., "Microstructure and hydrogenation kinetics of Mg_2Ni-based alloys with addition of Nd, Zn and Ti," Transactions of Nonferrous Metals Society of China, vol. 23, no. 12, pp. 3677–3684, 2013. ··

5. W. Oelerich, T. Klassen, and R. Bormann, "Comparison of the catalytic effects of V, V_2O_5, VN, and VC on the hydrogen sorption of nanocrystalline Mg," Journal of Alloys and Compounds, vol. 322, no. 1-2, pp. L5–L9, 2001. ··

6. M. A. Salam, S. Sufian, and T. Murugesan, "Catalytic hydrogen adsorption of nano-crystalline hydrotalcite derived mixed oxides," Chemical Engineering Research and Design, vol. 91, no. 12, pp. 2639–2647, 2013. ··

7. M. A. Salam, Y. Lwin, and S. Sufian, "Synthesis of nano-structured Ni-Co-Al hydrotalcites and derived mixed oxides," Advanced Materials Research, vol. 626, pp. 173–177, 2013. ··

8. S. Kang, S. Karthikeyan, and J. Y. Lee, "Enhancement of the hydrogen storage capacity of $Mg(AlH_4)_2$ by excess electrons: a DFT study," Physical Chemistry Chemical Physics, vol. 15, no. 4, pp. 1216–1221, 2013. ··

9. Y. Liu, L. P. Meng, S. J. Zheng, and S. W. Zhang, "The DFT studies on a novel hydrogen storage material $Mg_{12}Ni_{6-x}Cr_x(x=0,1)$," Applied Mechanics and Materials, vol. 457-458, pp. 181–184, 2014. ··

10. M. A. Salam, S. Sufian, and Y. Lwin, "Hydrogen adsorption study on mixed oxides using the density functional theory," Journal of Physics and Chemistry of Solids, vol. 74, no. 4, pp. 558–564, 2013. ··

11. K. W. Wiitala, T. R. Hoye, and C. J. Cramer, "Hybrid density functional methods empirically optimized for the computation of ^{13}C and ^{1}H chemical shifts in chloroform solution," Journal of Chemical Theory and Computation, vol. 2, no. 4, pp. 1085–1092, 2006. ··

12. D. Y. Chang, S. Y. Bong, S. N. Young, and S. B. Jong, "Hydriding properties of Mg-xNi alloys with different microstructures," Catalysis Today, vol. 120, pp. 276–280, 2007.

13. M. A. Salam, S. Sufian, and T. Murugesan, "Characterization of nano-crystalline Mg-Ni-Al hydrotalcite derived mixed oxides as

hydrogen adsorbent," Materials Chemistry and Physics, vol. 142, no. 1, pp. 213–219, 2013. ··

14. W. J. Hehre, L. Radom, P. V. R. Schleyer, and J. Pople, AB Initio Molecular Orbital Theory, John Wiley & Sons, New York, NY, USA, 1987.

15. M. J. Frisch, G. W. Trucks, H. B. Schlegel, et al., Gaussian 09W, Gaussian Inc, 2009.

16. R. G. Parr and R. G. Pearson, "Absolute hardness: companion parameter to absolute electronegativity," Journal of the American Chemical Society, vol. 105, no. 26, pp. 7512–7516, 1983. ··

17. R. G. Pearson, Chemical Hardness: Applications from Molecules to Solids, Wiley-VCH, Weinheim, Germany, 1997.

18. M. V. Putz, "Electronegativity: quantum observable," International Journal of Quantum Chemistry, vol. 109, no. 4, pp. 733–738, 2009. ··

19. K. M. Thomas, "Hydrogen adsorption and storage on porous materials," Catalysis Today, vol. 120, no. 3-4, pp. 389–398, 2007. ··

20. Ochterski, Thermochemistry in Gaussian, Gaussian, Inc, Wallingford, Conn, USA, 2000.

21. W. H. Zachariasen, C. E. Holley, and J. F. Stamper Jr., "Neutron diffraction study of magnesium deuteride," Acta Crystallographica, vol. 16, pp. 352–353, 1963. · View at Google Scholar

22. J. Zou, H. Sun, X. Zeng, G. Ji, and W. Ding, "Preparation and hydrogen storage properties of Mg-rich Mg-Ni ultrafine particles," Journal of Nanomaterials, vol. 2012, Article ID 592147, 10 pages, 2012. ··

23. O. Friedrichs, L. Kolodziejczyk, J. C. Sánchez-López, C. López-Cartés, and A. Fernández, "Synthesis of nanocrystalline MgH_2 powder by gas-phase condensation and in situ hydridation: TEM, XPS and XRD study," Journal of Alloys and Compounds, vol. 434-435, pp. 721–724, 2007. ··

24. R. J. Sibley and R. A. Alberty, Physical Chemistry, John Wiley & Sons, New York, NY, USA, 3rd edition, 2001.

25. X. Wang and L. Andrews, "Infrared spectra of magnesium hydride molecules, complexes, and solid magnesium dihydride," Journal

of Physical Chemistry A, vol. 108, no. 52, pp. 11511–11520, 2004. ··

26. E. N. Koukaras, A. D. Zdetsis, and M. M. Sigalas, "Ab initio study of magnesium and magnesium hydride nanoclusters and nanocrystals: examining optimal structures and compositions for efficient hydrogen storage," Journal of the American Chemical Society, vol. 134, no. 38, pp. 15914–15922, 2012.··

27. I. Swart, A. Filicki, B. Redlich, G. Meijer, B. M. Weckhuysen, and F. M. F. de Groot, "Molecular adsorption of H_2 on small cationic nickel clusters," Journal of the American Chemical Society, vol. 192, pp. 2516–2520, 2012.

28. Q. Zheng, Y. Pivak, L. P. A. Mooij et al., "EXAFS investigation of the destabilization of the Mg-Ni-Ti (H) system," International Journal of Hydrogen Energy, vol. 37, no. 5, pp. 4161–4169, 2012. ··

Microlayered Composite Materials on Basis of Copper, Refractory, Rare-Earth Metals, and Carbon for Electrical Contacts and Electrodes

Victor Volodymyrovych Bukhanovsky[1], Mykola Petrovych Rudnytsky[1], and Ilija Mamuzich[2]

[1]Department of Computational and Experimental Analysis for Structural Strength, Pisarenko Institute for Problems of Strength, National Academy of Sciences of Ukraine, Kyiv, Ukraine
[2]Faculty of Metallurgy, University of Zagreb, Zagreb, Croatia

ABSTRACT

A technology for obtaining microlayered composite materials of Cu-Zr-Y-Mo, Cu-Zr-Y-Cr, Cu-ZrY-W and Cu-Zr-Y-C systems by means of high-speed electron-beam evaporation-condensation, structure, electrical, and mechanical properties at ambient and elevated temperatures is developed.

INTRODUCTION

Composite materials (CMs) based on copper, refractory metals (tungsten, molybdenum and chromium) and carbon, which exhibit a number of unique physical, mechanical, chemical, technological and performance characteristics, were developed in the second half of the last century. Due to their specific structure and chemical composition they combine high heat-resistance, hardness, erosion and electric-erosion resistance, resistance against micro-welding typical of refractory metals, and high thermal and electrical conductivity, low contact resistance typical of copper and carbon. In addition, these materials are easy to process. Up to now, CMs of this type had been obtained by the methods of powder metallurgy. One of traditional powder metallurgy manufacturing processes is to press the refractory metals, sinter the pressed porous compact at a high temperature, and infiltrate with copper. Other traditional manufacturing process is hot-pressing of powder mix of refractory metals or carbon and copper. All this is done under very closely controlled conditions. Despite many years of experience with these materials, studies of their properties proceed within the system concerned in connection with the improved technical possibilities to control the material composition, dispersion and distribution of the refractory component in the product volume [1]-[7] .

CMs based on copper, refractory metals and carbon are used as electrical, structural and special purpose materials for the production of parts and structural elements to be in service under conditions of high temperatures, erosive and electroerosive wear, and other extreme operational factors. The composites of the Cu-W, Cu-Mo, Cu-Cr and Cu-C systems concerned found the widest application in electrical engineering as materials for contact parts of high-current switchgears, current-collecting devices, and electrodes for contact welding. The electrical and thermal conductivity of refractory metals, which is lower as compared to that of pure copper, calls for the creation of new technologies for producing combined products these composites can be used in such products only as coatings for contact parts of electrical switchgears that change their functional properties appreciably. In using this technology, difficulties occur due to the need in checking the quality of the boundary between the copper body and the composite coating. In a number of cases, these difficulties can be avoided, and a separate production of coatings and contact parts can be discontinued due to the technology that allows producing layered gradient composites and coatings based on them directly on the working surface of contact parts.

The production of such materials became possible after the creation of electron-beam facilities for high-speed evaporation of composite components from separate water-cooled crucibles with a subsequent layer-by-layer condensation of the regulated vapor flows on the revolving metallic substrate [8] -[17] . The apparent advantages of the electron-beam technology, which allowed obtaining the new-generation composite materials for electrical contacts, include:

- the possibility to mix at the atomic and molecular levels the vapor flows of substances that do not dissolve well in each other, and obtain composite materials and coatings (facing layers) with the assigned structure, chemical composition, physical, mechanical and performance characteristics, which cannot be yielded by other methods;

- simplicity and efficiency as compared to the methods of powder metallurgy, as the material is formed over one technological cycle;

- the possibility to create gradient structures by varying the deposition rate of the components being evaporated in the course of the technological process;

- the possibility to obtain layered composite materials, which is practically impossible to achieve using traditional methods;

- ecological purity, as this technology eliminates all atmospheric emissions.

MATERIAL, TREATMENT, AND TESTING

The objects for the investigation were the microlayered CMs of the Cu-Zr-Y-Mo, Cu-Zr-Y-Cr, Cu-Zr-Y-W and Cu-Zr-Y-C systems for electrical contacts, which were formed by the electron-beam technology at Gekont (Eltechmash) Science & Technology Company, Ukraine [8]-[10] . The electrical engineering materials condensed from the vapor phase were obtained on the L-5 electron-beam facility designed at the Paton Welding Institute of the National Academy of Sciences of Ukraine and upgraded at Gekont Science & Technology Company to suit our task. The scheme and surface appearance of the facility is given in Figure 1 and Figure 2.

The facility represents technological work chamber 1. The side wall of the work chamber has gun chamber 2 attached to it, which contains electron-beam heaters 3, 4, 5 and 6. The vacuum system, which comprises two fore pumps, two booster pumps and two high-vacuum units, serves to provide dynamic vacuum in the facility chambers for evaporation and condensation of the initial materials.

On the upper flange of the work chamber 1 there is mechanism 15 that rotates substrate 14. Substrate made in the form of a steel disk 1000 mm in diameter and 20 mm in width. The substrate surface which was presented to the crucibles and on which the condensation occurred was polished to achieve N8-N9 surface finishes. The design of rotation mechanism allows it to be operated for a long time without destroying vacuum at a temperature of 870 ± 50 K.

The substrate fixed to rotation rod 7 was heated to the assigned temperature by 40 kW electron-beam heaters 5 and 6. The initial materials were heated to evaporation by 100 kW electron-beam heaters 3 and 4. All heaters have independent cathode-glow and electron-beam controls.

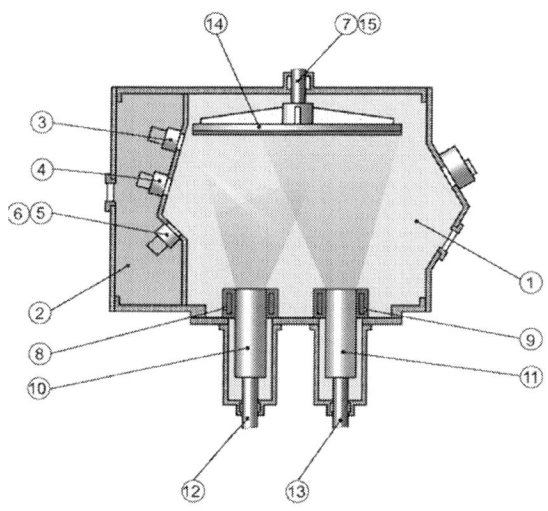

Figure 1: Scheme of the electron-beam facility. Designations: 1: work chamber; 2: gun chamber; 3-6: electron-beam heaters; 7: substrate rotation rod; 8: crucible for evaporation of copper; 9: crucible for evaporation of refractory metals or carbon; 10, 11: ingots of copper and refractory metals or carbon, respectively; 12, 13: mechanisms for supplying ingots in the vapor flow zone; 14: steel substrate for condensation of copper and refractory metals or carbon vapor flows; 15: substrate rotation mechanism.

Figure 2: Surface appearance of industrial electron-beam facility L-5.

The evaporation unit has two water-cooled copper crucibles 8, 9 of diameter 100 and 70 mm for evaporation of copper ingot 10 of diameter 98.5 ± 0.1 mm and length up to 500 mm and refractory metals (or carbon) ingot 11 of diameter 68.5 ± 0.1 mm and length 300 - 320 mm long to be evaporated, and mechanisms 12, 13 that allow the ingots to be put in the evaporation zone.

In the present investigation, we obtained the CMs based on copper, refractory and rare-earth metals by means of copper, zirconium, yttrium and tungsten (molybdenum, chromium or carbon) evaporation from separate crucibles followed by their precipitation on a rotated steel substrate coated with a separating layer of calcium fluoride. The initial materials were M0 grade copper ingots with 100 mm diameter, MChVP grade molybdenum (Kh99 grade chromium, VA grade tungsten or MPG-7 carbon) ingots with 70 mm diameter as well as toughpitch zirconium and yttrium batches.

The copper matrix was alloyed with zirconium and yttrium as follows. Two batches of these components, 135 g each, were put on the surface of the copper ingot. On reaching a vacuum to $(1.3 - 4.0) \times 10^{-3}$ Pa in the work chamber, we performed the electron-beam heating of the substrate, on which vapors were to be condensed, up to the temperature 950 ± 15 K. Simultaneously, we heated the surface of the copper ingot, making the constituents (zirconium, yttrium, copper) lying on it melt at a current of 1.15 - 1.3 A. The melt pool became homogeneous after 15 - 20 minutes of heating. At the production stage the evaporation of copper ingots was performed at a beam current of 2.6 - 2.8 A under acceleration voltage 20 kV. The total content of zirconium, yttrium, and their oxides in composites was no more than 0.8 wt.% in all cases.

Tungsten (molybdenum, chromium or carbon) was evaporated from the other crucible. Considering very high melting temperature of carbon and the difficulty of its transformation into a vaporous state the original electron-beam technology of carbon evaporation through molten tungsten mediator was designed by the Gekont Science & Technology Company. By varying the beam current in the range from 1.7 to 3.8 A under the direct acceleration voltage 20 kV, one can readily regulate the evaporation rate of refractory components and its concentration in the composite in wide ranges. Optimal content of refractory component in composite, which yields the most beneficial combination of their electric, mechanical, chemical, and operating characteristics, depends on material functionality and is determined experimentally. Generally for tungsten it lies between 5 - 25 mas.%, for molybdenum between 2.5 - 12.0 mas.%, for chromium between 30 - 40 mas.%, and for carbon between 1.2 to 7.5 vol.%.

By maintaining the substrate temperature in the range from 935 to 965 K, we prevented the re-evaporation of copper from the surface of the condensed material. The resulting condensed materials was a ~3···5 mm thick plate.

At the end of the technological process condensed composite material was separated from the substrate and annealed in the vacuum furnace at 1170 K for 3 h in order to relieve internal stresses, make the structure stable and enhance the material ductility.

At present time commercial and experimental-industrial condensed CMs of the Cu-Zr-Y-Mo, Cu-Zr-Y-Cr, Cu-Zr-Y-W, and Cu-Zr-Y-C systems for electrical contacts and electrodes are manufacturing in Ukraine by the Gekont Science & Technology Company the in the form of sheets by 3···5 mm thick. Generally, these sheet materials are used as operating floors of copper electrical contacts and electrodes attached to them by brazing.

In this study, we used the research techniques that include the macro-and microstructure analysis using optical and scanning electron microscopy, electrical resistance methods, and mechanical tensile tests at room and high temperatures, measurement of hot hardness.

The structure, chemical composition, electro physical, tribotechnical and mechanical properties of the condensed materials were investigated on specimens cut from the plates. Specimen dimensions were set in accordance with the requirements for specimens to be used for studying the material structure, electrical conductivity, microhardness, tribotechnical and mechanical characteristics during tensile tests. The content of tungsten, molybdenum, chromium, carbon and copper in composites was determined by the mortar method (volumetric analysis).

The structure of the composite materials was investigated by optical and scanning electron microscopy using the "Neophot-2" optical microscope and the "Superprobe 733" raster electron microscope manufactured by "Jeol". Specimens for metallographic analysis were prepared using chemical etching in a 40% hydrochloric acid solution and ion etching in a glow discharge. We studied the specimen surface and cross section perpendicular to the substrate, on which vapors were to be condensed.

The mechanical characteristics were determined at ambient temperature (in air) and in the range 370 - 1070 K (at a pressure no more than 0.7 mPa) from the results of mechanical tensile tests of standard flat fivefold proportional specimens with a gauge length of 15 mm, width of 3 mm and thickness of 1.2 - 2.0 mm using a 1246-R unit [18] according to ISO 6892 [19] and ISO 783 [20] , respectively. The specimens were cut from 1.2 - 2.0 mm thick composite materials. We tested from 3 to 5 specimens at each temperature. The deformation rate was 2 mm/min, which corresponded to a relative strain rate of ~2.2 × 10^{-3} s^{-1}. During the tests deformation diagrams were recorded to determine the proof strength $R_{p0,2}$, the ultimate strength R_m, the percentage elongation after fracture A, and the percentage reduction of cross-sectional area Z.

Hardness of the CMs was measured in the temperature range from 20 to 1070 K by Vickers indentation in the plane parallel to the surface of condensation. The pyramidal point was made of a synthetic corundum single crystal. Indentation loads were 10 N. The tests were carried out at a pressure no more than 0.7 mPa on a UVT-2 unit [21] [22] according to the DSTU 2434-94 [23] .

RESULTS AND DISCUSSION

Numerous properties requirements which a material for electrical contacts should meet are highly inconsistent and are mainly dictated by their operating conditions. To meet these requirements, the contacts should be made from a material possessing a certain complex of physical, structural, mechanical, and electrical contact properties. In particular, these properties include high electrical and thermal conductivity, melting temperature, high critical current strengths and voltages, strength characteristics, low contact resistance, high electrical erosion and corrosion resistance, volatility of its oxides, ease of processing, relatively low prime cost and environmental compatibility. At present, none of the known electrical contact materials possess the full set of the above properties.

By applying the layer-by-layer condensation of low-alloy copper and refractory metals (tungsten, molybdenum or chromium) or carbon and moving the substrate out of the zone of vapor flow while it is rotating, it is possible to perform tempering from vapor and obtain materials with the structural elements typical of nanomaterials. Investigated composites are characterized by a specific microlayer structure with alternating layers of Cu-Zr-Y low alloy and refractory metals from 0.1 to 0.4 μm thick (Figure 3(a) and Figure 3(b)). The grain size of copper in the composite is 0.1 - 0.3 μm and that of tungsten (molybdenum or chromium) from 0.01 - 0.02 μm. Composites Cu-Zr-Y-C has layered structure with cuprum grain size 0.1 - 0.3 μm and disperse particles of carbon which mean size does not exceed 200 Å (Figure 3(c) and Figure 3(d)).

Commercial composite materials by Cu-Zr-Y-Mo system for electrical contacts are produced serially by the Gekont Science & Technology Company in three modifications: MDK-1, MDK-2 and MDK-3 with the molybdenum content of 2.5 - 5.0, 5.1 - 8.0, and 8.1 - 12.0 wt. %, respectively. Standard values of physical and mechanical characteristics of MDK grade composite materials and rated currents for contacts produced on their basis are given in Table 1.

In operation, the materials of contacting pairs in high-current switchgears are subjected not only to intensive corrosion and electrical erosion but also to mechanical loads at elevated temperatures. Therefore, studies on their mechanical properties over operating temperature ranges are of definite scientific and practical interest. Table 2 lists the mechanical characteristics of the most promising composite material MDK-3 in the temperature range from 290 to 1070 K. This composite material exhibit a unique complex of mechanical characteristics combining high hardness and strength with a satisfactory plasticity in the whole temperature range studied. The strength of this composite material in 2 or 4 times higher than that of known Cu-Al$_2$O$_3$ and Cu-BeO powder composites and cast copper super alloys

The electrical conductivity of MDK composite materials by Cu-Zr-Y-Mo system ranges from 65% to 75% of that of copper, which is almost twice as much as that of all known Cu-Mo powder compositions. Their maximum magnitude of transferred current (up to 4000 A) is 2.5 times higher than that of silver. MDK composite materials significantly surpass all existing electrical engineering materials in radiation resistance, thermal stability, and wear resistance. They exhibit high thermal conductivity, do not maintain the arc, and are more corrosion-resistant and durable than silver [10].

Additional alloying of the copper matrix with yttrium and zirconium allowed us to modify greatly the composition of the oxide film appearing in operation when the condensed material is in an oxidizing medium. Unlike Mo-Cu powder composites where the oxide films consist of $CuO \times MoO_3$ and $3CuO \times 2MoO_3$ compounds, the films of condensed materials are formed on the basis of complex spinels, such $CuMoY_2O_7$ and $CuMoZrY_2O_9$. In such oxide films no polymorphic transformations occur and they are distinguished by a high electrical conductivity and adhesion to the base material.

Electrical contacts made of Cu-Zr-Y-Mo microlayer CMs exhibit high thermal and electrical corrosion resistance. They are unweldable and meet all increasing demands for the reliability and service life of high-current switchgear (Figure 4(a)).

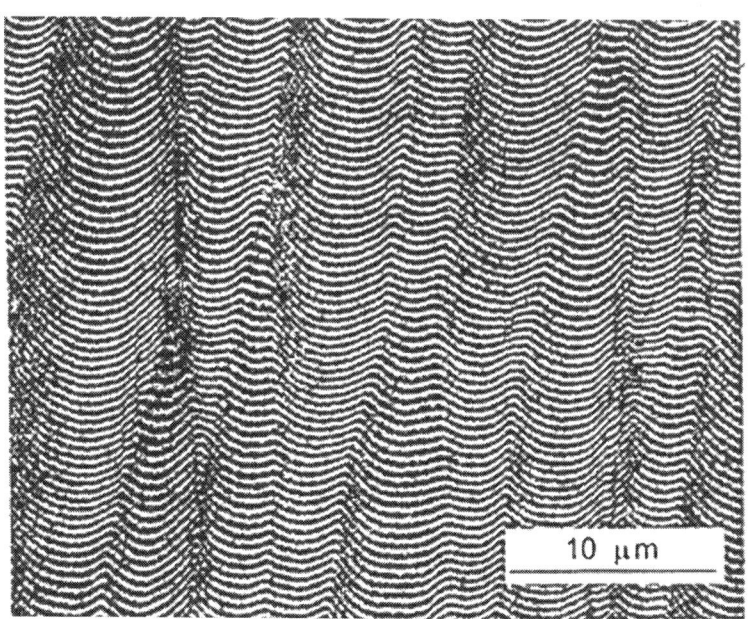

10 μm

(a)

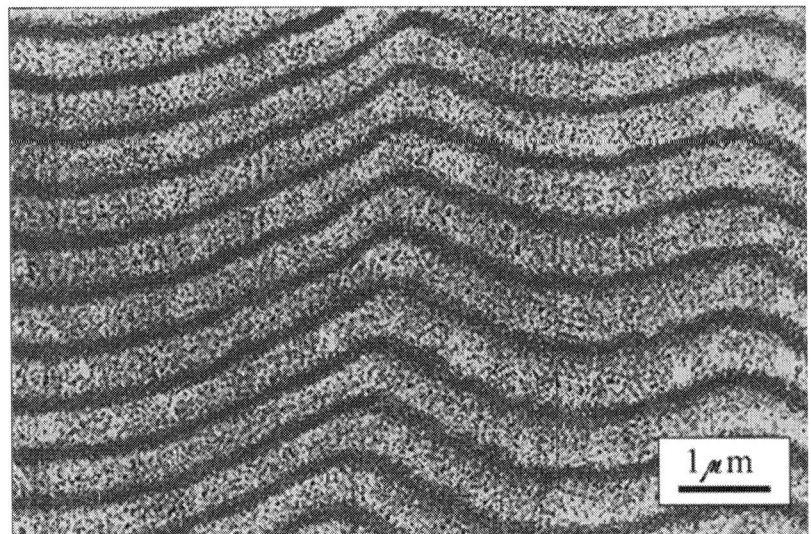

(b)

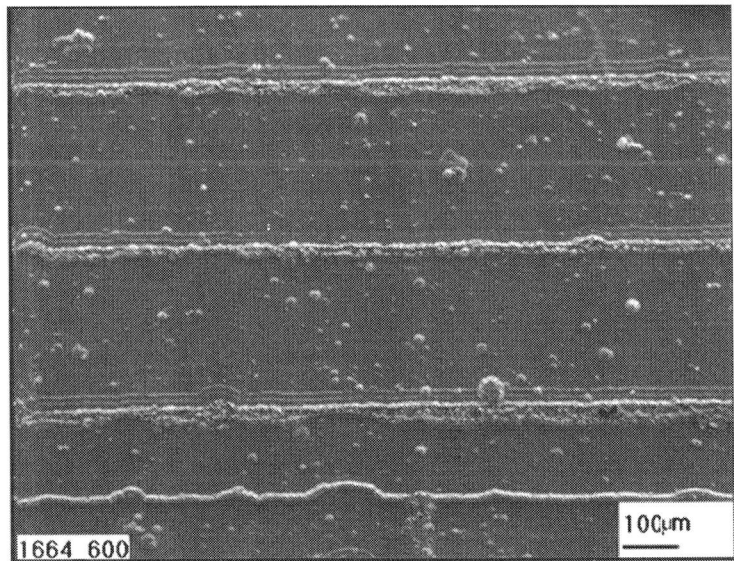

(c)

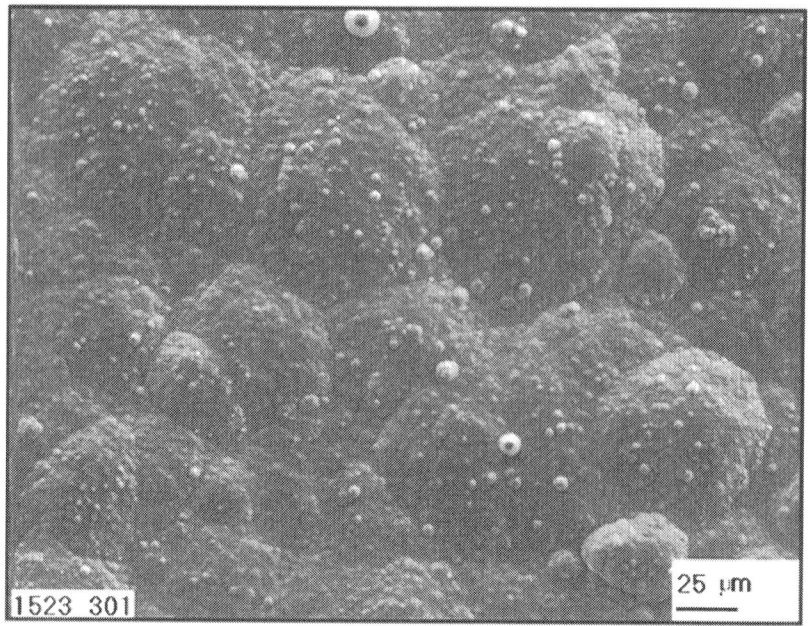

(d)

Figure 3: Microstructure of Cu-Zr-Y-Mo (a, b) and Cu-Zr-Y-C (c), (d) composite materials: (a)-(c): micro-layer structure of the composites, observed after ion etching; d: composite surface microstructure (without ion etching). Scanning electron microscopy.

Table 1: Standard values of physical, mechanical, and performance characteristics of mdk materials

Material type	Molybdenum content, wt.%	Density, g/cm²	Vickers hardness, MPa	Electrical resistivity, Ohm·mm²/m	Rated current of contacts, A
MDK-1	2.5 - 5.0	8.98 - 9.00	1000 - 1500	0.021 - 0.022	up to 10
MDK-2	5.1 - 8.0	9.00 - 9.05	1500 - 1650	0.022 - 0.024	up to 100
MDK-3	8.1 - 12.0	9.05 - 9.10	1650 -1750	0.024 - 0.028	up to 100

Table 2: Hardness, strength, and plasticity of MDK-3 composite material in the temperature range from 290 to 1070 K

T, K	HV, MPa	R_m, MPa	$R_{p0,2}$, MPa	A, %	Z, %
290	1800	645	595	8.7	37.6
370	1350	570	535	8.8	18.3
470	1065	475	440	12.4	33.1
570	775	350	330	1.8	7.0
670	575	290	265	7.9	13.5
770	414	205	185	8.8	12.8
870	305	140	130	13.1	9.1
970	215	82	73	17.1	16.6
1070	150	45	39	21.7	33.7

Nowadays composites based on copper and chromium is the most effective electrical contact materials for interrupter contacts in vacuum circuit breakers. Experimental-industrial layered composite materials by Cu-ZrY-Cr system with chromium content of 30.0 - 40.0 mas. % is produced in Ukraine by the Gekont (Eltechmash) Science & Technology Company. Electrical resistivity of this composites does not exceed 2.5×10^{-2} Ohm·mm²/m. Mechanical characteristics of the Cu-Zr-Y-Cr composite material with average chromium content of 40.0 ± 0.5 mas.% in the temperature range from 290 to 1070 K are given in Table 3.

(a)

(b)

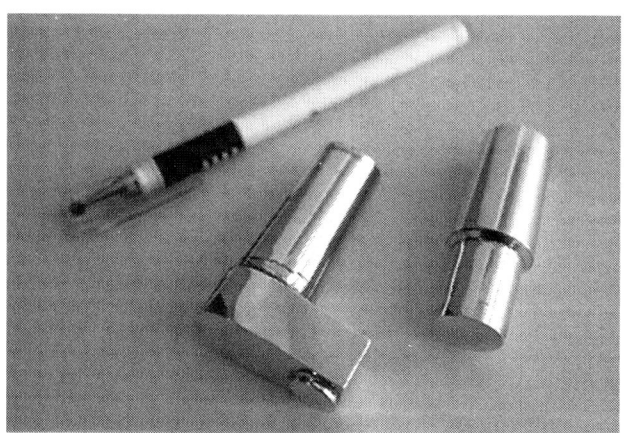

(c)

(d)

Figure 4: Production samples of electrical contacts and electrodes made on basis of Cu-Zr-Y-Mo (a), Cu-Zr-Y-Cr (b), Cu-Zr-Y-W (c), and Cu-Zr-Y-C (d) microlayer CMs: (a) gamma-type interrupting electrical contacts for high-current switchgears; (b) cylindrical interrupting contacts for vacuum circuit breakers; (c) electrodes for contact welding; (d) sliding contact for current-collecting devices.

Table 3: Hardness, strength, and plasticity of Cu-Zr-Y-Cr composite material with average chromium content of 40.0 ± 0.5 mas. % in the temperature range from 290 to 1070 K

T, K	HV, MPa	R_m, MPa	$R_{p\ 0,2}$ MPa	A, %	Z, %
290	1690	362	257	10.5	28.8
370	1435	310	220	27.3	55.5
470	1230	220	160	38.6	64.5
570	995	157	116	45.0	65.0
670	790	125	93	48.5	56.0
770	620	88	70	39.0	45.4

870	435	62	53	25.7	38.3
970	245	38	35	23.3	37.5
1070	138	21	19	22.0	35.0

Electrical interrupting contacts for vacuum circuit breakers made of Cu-Zr-Y-Cr microlayer composite materials exhibit high thermal and electrical corrosion resistance. They are unweldable and meet all increasing demands for the reliability and service life of vacuum circuit breakers (Figure 4(b)).

Owing to the highest hardness and strength together with satisfactory plasticity at higher temperatures (Table 4) condensed composites of Cu-Zr-Y-W system are very promising as materials for operating floors of electrodes in contact welding (Figure 4(c)). The electrical resistance of these composites depends on tungsten content and does not exceed 3.05×10^{-2} Ohm·mm^2/m (for CMs with 25 mas. % W).

Condensed layered CMs of Cu-Zr-Y-C system with high tribotechnical characteristics are used for manufacturing sliding contacts for current-collecting devices of electric transport (Figure 4(d)). The coefficient of kinetic friction of Cu-Zr-Y-C composites with 4.0 - 7.0 vol.% C together with copper contact wire is 3 - 4 fold lower than that of the tough-pitch copper. Mechanical characteristics of the Cu-Zr-Y-C composite material with average carbon content of 5.0 ± 0.2 vol. % in the temperature range from 290 to 870 K are given in Table 5.

SUMMARY

- A manufacturing technology for obtaining condensed microlayered composite materials of the Cu-ZrY-Mo, Cu-Zr-Y-Cr, Cu-Zr-Y-W, and Cu-Zr-Y-C systems by means of high-speed electron-beam evaporationcondensation is developed. The CMs components are evaporated from separate crucibles and then precipitated on a rotating steel substrate. Condensed composites with a thickness of 3 - 5 mm and regulated content of refractory components, which are very promising as materials for electrical contacts and electrodes, are obtained industrially.

Table 4: Hardness, strength, and plasticity of Cu-Zr-Y-W composite material with average tungsten content of 25.0 ± 0.5 mas. % in the temperature range from 290 to 1070 K

T, K	HV, MPa	R_m, MPa	$R_{p0,2}$, MPa	A, %	Z, %
290	1950	680	650	13.5	27.0
370	1660	600	565	14.8	34.3
470	1420	510	465	16.4	36.1
570	1145	375	350	17.8	37.5
670	910	310	285	19.5	41.5
770	720	260	185	20.4	40.8
870	505	185	170	17.5	35.5
970	360	132	117	18.0	34.5
1070	250	75	65	20.7	32.5

Table 5: Hardness, strength, and plasticity of Cu-Zr-Y-C composite material with average carbon content of 5.0 ± 0.2 vol.% in the temperature range from 290 to 870 K

T, K	HV, MPa	R_m, MPa	$R_{p0,2}$, MPa	A, %	Z, %
290	951	257	225	8.5	28.2
370	724	213	186	6.7	24.6
470	571	167	153	4.6	22.0
570	381	127	117	4.5	20.2
670	290	104	96	6.0	18.3
770	177	65	59	6.6	17.4
870	127	37	34	8.2	17.0

- Structure, electrical conductivity, hardness, tribotechnical and mechanical properties in tension of the microlayer CMs for electrical contacts and electrodes obtained by means of high-speed electron-beam evaporation-condensation were investigated at ambient and elevated temperatures.

- The practical use of condensed CMs as materials for electrical applications will be based on a preliminary study of the operating conditions of the electrical contact or electrode, choice of chemical composition for a future combined product, and a detailed study of its production technology, which is determined by the required functional properties of the electrical switching devices.

- Many years of series production and practical use of condensed microlayered CMs for electric contacts at Gekont Science & Technology Company in Ukraine gave grounds to believe that the offered materials would be promising enough for wide commercial applications.

ACKNOWLEDGEMENTS

Authors express their thanks to president of Gekont Science & Technology Company professor N.I. Grechanyuk and professor of Frantsevich Institute for Problems of Materials Science of National Academy of Sciences of Ukraine R.V. Minakova for the provided composite materials and help rendered during metallographic and electrophysical investigations.

REFERENCES

1.   Gnesin, G.G. (1981) Sintered Materials for Electrical and Electronic Applications. Handbook, Metallurgy, Moscow.

2.   Rakhovsky, V.I., Levchenko, G.V. and Teodorovich, O.K. (1966) Interrupting Contacts of Electric Appliances, Energy. Leningrad, Moscow.

3.   Janković Ilić, D., Fiscina, J., González-Oliver, C.J.R., Ilić, N. and Mücklich, F. (2008) Electrical and Elastic Properties of Cu-W Graded Material Produced by Vibro Compaction. Journal of Materials Science, 43, 6777-6783. http://dx.doi.org/10.1007/s10853-008-2941-2

4.   Li, Y., Qu, X., Zheng, Z., Lei, C., Zou, Z. and Yu, S. (2003) Properties of W-Cu Composite Powder Produced by a Thermo-Mechanical Method. International Journal of Refractory Metals

and Hard Materials, 21, 259-264. http://dx.doi.org/10.1016/j.ijrmhm.2003.08.001

5.   Chen, G., Jiang, L., Wu, G., Zhu, D. and Xiu, Z. (2007) Fabrication and Characterization of High Dense Mo-Cu Composites for Electronic Packaging Applications. Transactions of Nonferrous Metals Society of China, 17, 580-583.

6.   Müller, R. (1988) Arc-Melted Cu-Cr Alloy as Contact Materials for Vacuum Interrupters, Siemens Forsch. Entwicklungsber, 17-33, 105-111.

7.   Kim, M., Doh, J., Park, J. and Jung, J. (2001) Microstructural Evolution in the Cr-Cu Electric Contact Alloys during Liquid Phase Sintering. Proceedings of the 15th International Plansee Seminar, Plansee Holding AG, Reutte, 1, 29-43.

8.   Grechanyuk, N., Mamuzić, I. and Shpak, P. (2002) Modern Electron-Beam Technologies of Melting and Evaporation of Materials in Vacuum, Used by Gekont Company, Ukraine. Metalurgija, 41, 125-128.

9.   Grechanyuk, N.I., Osokin, V.A., Grechanyuk, I.N. and Minakova, R.V. (2005) Vapour-Phase Condensed Materials Based on Copper and Tungsten for Electrical Contacts. Structure, Properties, Technology. Current Situation and Prospects for Using Electron Beam Physical Vapour Deposition Technique to Obtain Materials for Electrical Contacts, First Announcement. Modern Electrometallurgy, 2, 28-35.

10. Grechanyuk, N.I., Osokin, V.A., Grechanyuk, I.N. and Minakova, R.V. (2006) Basics of Electron Beam Physical Vapour Deposition Technique Used to Obtain Materials for Electrical Contacts. Structure and Properties of Electrical Contacts, Second Announcement. Modern Electrometallurgy, 2, 9-19.

11. Bukhanovskii, V.V., Rudnitsky, N.P. and Mamuzić, I. (2006) Effect of Chemical Composition and Heat Treatment on Mechanical Characteristics of Microlayer Composite Materials of the Cu-Zr-Y-Mo System in Wide Temperature Range. Metals Science and Heat Treatment, 48, 113-117. http://dx.doi.org/10.1007/s11041-006-0053-7

12. Grechanyuk, N.I., Mamuzić, I. and Bukhanovsky, V.V. (2007) Production Technology and Physical, Mechanical, and

Performance Characteristics of Cu-Zr-Y-Mo Finely-Dispersed Microlayer Composite Materials. Metalurgija, 46, 93- 96.

13. Bukhanovskii, V.V., Mamuzić, I. and Rudnitsky, N.P. (2008) The Effect of Temperature on Mechanical Characteristics of Copper-Carbonic Composite. Kovove Materialy (Metallic Materials), 46, 33-37.

14. Grechanyuk, N.I., Mamuzić, I. and Minakova, R.V. (2008) Peculiarities of the Structure, ITS Deformation and Destruction of Condensed Cu-Mo-Zr-Y Composite Material of Commercial Purity. Metalurgija, 47, 99-102.

15. Bukhanovskii, V.V., Rudnitsky, N.P. and Mamuzić, I. (2009) The Effect of Temperature on Mechanical Characteristics of Copper-Chromium Composite. Materials Science and Technology, 25, 1057-1061. http://dx.doi.org/10.1179/174328408X365829

16. Bukhanovskii, V.V., Rudnitsky, N.P., Mamuzić, I., Minakova, R.V. and Grechanyuk, N.I. (2009) Effect of Composition and Process Factors on the Structure, Mechanical Properties, and Fracture Behavior of a Composite Material of Copper-Chromium System. Metals Science and Heat Treatment, 51, 388-393. http://dx.doi.org/10.1007/s11041-009-9168-y

17. Bukhanovskii, V.V., Rudnitsky, N.P., Kharchenko, V.V. and Mamuzić, I. (2010) Relationship between Hardness and Strength Characteristics of Cu-Cr Microlayer Composite Material at Elevated Temperatures. Strength of Materials, 42, 187-196.http://dx.doi.org/10.1007/s11223-010-9206-4

18. Kluev, V.V. (1982) Test Equipment. Handbook, Mashinostroenie, Moscow.

19. Metallic Materials (1998) Tensile Testing at Ambient Temperature.

20. Metallic materials (1999) Tensile Testing at Elevated Temperature.

21. Aleksyuk, M.M., Borisenko, V.A. and Krashchenko, V.P. (1980) Mechanical Tests at High Temperatures. Naukova Dumka, Kiev.

22. Borisenko, V.A. (1984) Hardness and Strength of Refractory Materials at High Temperatures. Naukova Dumka, Kiev.

23. Borisenko, V.A. and Oksametna, O.B. (1995) DSTU 2434-94 Method of Determining High-Temperature Hardness by Pyramidal and Bicylindrical Indentation. Kiev.

Low Velocity Impact and Creep-Strain Behaviour of Vinyl Ester Matrix Nanocomposites Based on Layered Silicate

A. I. Alateyah[1, 2], H. N. Dhakal[2], Z. Y. Zhang[2], and B. Aldousiri[3]

[1]Al Imam Mohammad Ibn Saud Islamic University (IMSIU), Riyadh 13318, Saudi Arabia

[2]Advanced Polymer and Composites (APC) Research Group, School of Engineering, University of Portsmouth, Portsmouth PO1 3DJ, UK

[3]Department of Power and Desalination Plants, Ministry of Electricity, South Surra, 13001 Kuwait, Kuwait

ABSTRACT

The impact properties of neat vinyl ester and the nanocomposites were performed using a low velocity impact testing. The addition of layered silicate into the polymer matrix shows that an optimum range of nanoclay reinforcement in the vinyl ester matrix can produce enhanced load bearing and energy absorption capability compared to the neat matrix. In addition, the amount of microvoids in the nanocomposites structure influences the overall properties. Likewise, the influence of the clay addition into the neat polymer on the creep relaxation behaviour at 25°C and 60°C was studied. In both cases, the presence of the layered silicate remarkably improved the creep behaviour. The improvement of these properties can be assigned to the stiff fillers and the configurational linkage between the polymer and the layered silicate which are supported by scanning electron microscopy (SEM) and transmission electron microscopy (TEM) characterisations by showing a distinct change in surface morphology associated with improved impact toughness and creep response.

INTRODUCTION

Polymer materials are widely used in different research and industrial fields, owing to their advantageous properties such as light weight and manufacturing simplicity. However, certain of the polymer properties are inadequate unless they are enhanced through incorporation of fillers and various reinforcements leading to the formation of composite or nanocomposite materials [1]. For that reason and to overcome these downsides, suitable additives are utilised in a host of pristine polymers in order to improve their properties [2].

Polymers with various additives have been successfully reinforced to improve their properties, such as mechanical, thermal, and barrier properties [3, 4]. The presence of the particulate fillers often results in undesirable properties, such as brittleness and opacity. Also, the dispersion of the additive into the polymer is not homogenous [5].

Composite materials have a fairly new class of material nanocomposites that combine between fillers and matrix and in which at least one dimension of the dispersion particles is in the nanometer range. The use of nanocomposites prepared by a layered structure, like clays, has been the subject of elaborate research. However, the subject is experiencing resurgence, both in terms of academic research and industrial activity due to the numerous properties that nanocomposites stand to afford [6].

Polymer layered silicate nanocomposites usually provide more attractive enhancements to material properties than conventional composite materials [7–12]. The improvement of properties can be mechanical (strength, modulus, and hardness), thermal, or barrier properties [9, 13–18]. In addition, the polymer performance can be enhanced by the addition of layered silicate [19]. The percent of the nanofillers is usually less than 5 wt.% clay loading, which can enhance the engineering properties without sacrificing important properties such as optical and weight [20]. However, the impact properties are a big concern in the field of nanocomposites, where many studies noted a reduction of impact properties by the addition of layered silicate [21]. In some applications in hard working conditions, such as slide bearings, the impact properties are the main key to meet the requirements of these applications [22]. Thus, the study of the impact properties is fundamentally important for many polymers. In addition, another fundamental property is creep, which can be defined as the time dependent deformation of the materials that are subject to continuous load under the yielding stress of the materials. The deformation of creep can be either plastic or elastic, which may be nonrecoverable after the relaxation load. The creep can end up causing the material to have a structural failure, so the study of this property is important for different engineering designs [23].

Thermoset vinyl ester resin was used in this study due to its enhanced mechanical and chemical resistance properties. It is a middling choice between the epoxy and unsaturated polyester in terms of its properties and cost. However, the vinyl ester as a matrix has some drawbacks which include low resistance to crack propagation, brittleness, and shrinkage under polymerisation [24]. Thus, the primary objective of this study is to investigate the effect of the incorporation of the layered silicate into the vinyl ester matrix on the low velocity impact and creep-strain relaxation behaviour.

EXPERIMENTAL

Materials

The matrix material used in this study is vinyl ester (VE) resin. This material was purchased locally from Cathy Composites Portsmouth and commercially coded as AME 6000 T 35. The layered silicate that has been used is Cloisite 10A which is classified as a natural montmorillonite that is modified with a quaternary ammonium salt.

Sample Fabrication Process

Neat Vinyl Ester

In order to make neat vinyl ester panels, the vinyl ester was directly mixed with the curing agent (MEKP) (mix ratio 1.5%) and then was poured in a steel mould. The mould was closed and the composite panel was left to cure in a hydraulic press at a temperature of 55°C and at a compaction pressure of 1 MPa for 2 hours.

Nanocomposites

Prior to this process, the layered silicate was dried for 3 hours at 120°C in a fan assisted oven. The vinyl ester resin was mixed with various concentrations of nanoclay at room temperature using a mechanical mixer in an ultrasonic bath for 2 hours. A degassing process was applied in the mixture for 3-4 hours and then it was left overnight in order to get rid of the remaining air bubbles naturally. A curing agent (MEKP) was added to the mixture (1.5%). A Frekote mould release was utilised in order to easily remove the samples. The mould was closed and the composite panel was left to cure in a hydraulic press at a temperature of 55°C and at a compaction pressure of 1 MPa for 2 hours. A postcuring process for 3 hours at 80°C followed. The concentrations of the layered silicate were 0, 1, 2, 3, 4, and 5 wt.%.

Characterisation

Wide Angle X-Ray Diffraction (WAXD)

WAXD analysis on compression-moulded specimens was used to determine the clay intercalation and interlayer spacing utilising a Philips APD 1700 X-ray diffraction system with Cu K radiation ($\lambda = 1.542$ A) generated at 40 mA and 40 kV. The basal spacings (the d-spacing, in Angstroms, between layers) were calculated using Bragg's law.

Scanning Electron Microscopy (SEM)

The morphology of vinyl ester/nanocomposite systems was investigated in a Hitachi S4500 SEM working at an operating voltage of 8 kV. Block faces were prepared from each material, and then ultrathin sections (63 nm) were collected using a diamond knife in a Reichert Ultracut E ultramicrotome. Plasma etching was used to preferentially remove the vinyl ester matrix and leave the clay particles sitting proud of the surface. After adhering to SEM stubs, a thin layer of gold/palladium was applied to the specimens prior to examination in a Quanta 250 FEG SEM.

Energy Dispersive X-Ray Spectrometry (EDS)

The morphology of the VE/nanocomposite structure was further examined using a Jeol JSM 6060LV microscope working at an operating voltage of 8 kV. The degree of dispersion between the layered silicate and the vinyl ester matrix of the nanocomposite samples was measured using energy dispersive spectroscopy (EDS), a by-product of the back-scattered electrons off the specimen from the electron beam. By scanning the beam in television-like raster and showing the intensity of the selected sample, a map (image) of the distribution of elements can be produced.

Transmission Electron Microscopy (TEM)

TEM measurements on vinyl ester/nanocomposite systems were performed using a high-resolution transmission electron microscope (Phillips CM12 with an associated Gatan digital camera system). The same block faces used to produce the sections for SEM examination were also used for TEM.

Testing

Impact Test

The impact strength of the neat vinyl ester and the corresponding nanocomposite samples was determined by an instrumented falling weight impact tester (Zwick Roell, HIT230F). The annular hole diameter on the specimen fixture was approximately 4 cm. The specimen dimension utilised for the impact test was 60 mm × 60 mm × 6 mm. The total mass (kg), work capacity (J), and the height of release of the load (mm) were 23.11 kg, 25 J, and 110 mm, respectively. This process was carried out at room temperature. The energy absorption of the samples was calculated from the curve of the maximum force deformation. The incident energies were obtained from adjusting the drop height of the impactor and calculated using a typical energy equation:

$$E_i = mgh,$$

(1)

where E_i is incident impact energy, m is mass of the impactor, g is gravity, and h is height.

Creep Relaxation Behaviour

The creep test was carried out with the following parameters: 60 N loads, 25°C and 60°C, and 96 hrs period times. Neat polymer, 2 wt.%, and 4 wt.% nanocomposite samples were investigated. The specimens were prepared as 20 mm × 20 mm × 6 mm. This size sample was used in order to compare it with different mechanical tests such as tensile, flexural, and nanoindentation that have been already applied on the same geometry sample [6]. The initial and final strains of each sample were calculated from the strain-time curve.

RESULTS AND DISCUSSIONS

Characterisations of the Interlamellar Structure and Surface Morphology

Wide Angle X-Ray Diffraction (WAXD)

XRD values of the neat polymer and the corresponding nanocomposites can be seen in Table 1 and Figure 1. The Cloisite 10A represents 20° which indicates 0.443 nm basal distance. At 2 wt.% clay loading, the angle was shifted toward a lower angle compared to Cloisite 10A and showed 16.86° which indicated 0.525 nm, so the intercalation of the nanocomposite structure took place. The enhancement of the d-spacing of the layered silicate of 2 wt.% was 16% compared to the pristine clay. At 4 wt.%, the 2θ was much reduced compared to both neat clay and 2 wt.% and represents 13.84° which indicated an intercalated basal spacing of 0.640 nm. The enhancement of the interlayer spacing at 4 wt.% was about 45% compared to the d-spacing of base clay. This enhancement in d-spacing value at the 4 wt.% reinforced samples indicated that the nanocomposites structure was intercalated or partially well-dispersed. In addition, the increment of basal distance indicated a good dispersion level of the clay into the polymer matrix. After the addition of more clay (i.e., 5 wt.%), the 2θ value was 16.08° and represented 0.551 nm of layered silicate spacing which was reduced compared to 4 wt.%. The reduction of d-spacing of 5 wt.% may be attributed to the high viscosity of the mixture where insufficient mixing might have taken place, so agglomeration layers were observed in the nanocomposite structure. A clear relationship between the gallery distance and the level of distribution of the layered silicate in the matrix is proved by the 2θ values.

Table 1: XRD results obtained from different clay loading of nanocomposites

Sample number	2 values	The interlayer distances (nm)	d-spacing improvement %

Cloisite 10A	20.00	0.443	00.00
Vinyl ester + 2 wt.% clay	16.86	0.525	18.51
Vinyl ester + 4 wt.% clay	13.84	0.639	44.24
Vinyl ester + 5 wt.% clay	16.08	0.551	24.38

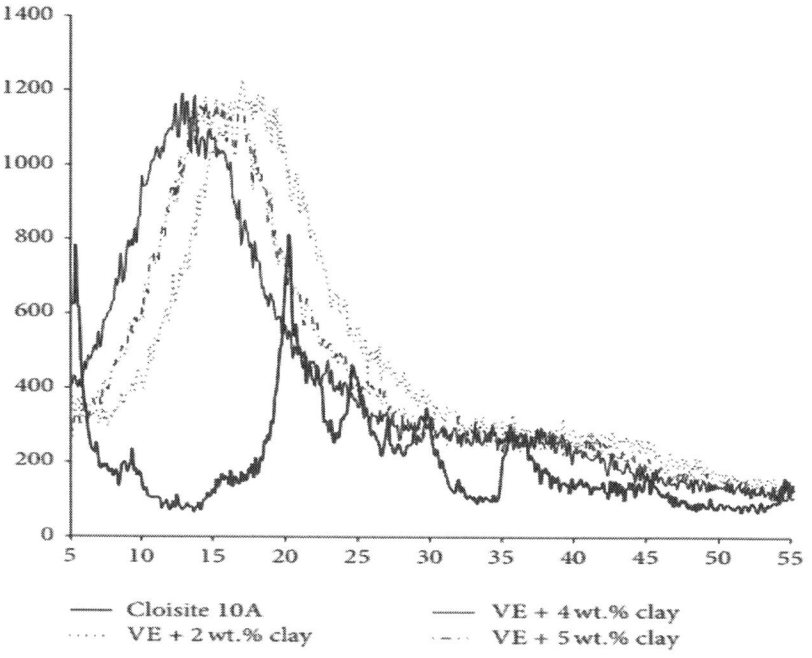

Figure 1: XRD results of neat polymer and the corresponding nanocomposites.

Scanning Electron Microscopy (SEM)

The SEM images in Figure 2 illustrate clearly the incorporation of the layered silicate through the vinyl ester for each of the three levels of loading. Because of the difference between the vinyl ester and the clay in scattering density, the large aggregation particles can be easily illustrated by SEM. As the selected images show below, the largest clay agglomerates are of a similar size for all three samples, being around 30 to 35 microns in size. However, their frequency increases with each larger clay volume fraction, as does the degree of infilling between them with smaller aggregation. It can be seen that the 2 wt.% clay loading exhibits uniform distribution layers throughout the polymer sea. Likewise, the well-dispersed clay within the vinyl ester matrix and unpronounced aggregation of layered silicate took place at higher amounts of clay such as 4 wt.%. The SEM image of 5 wt.% clay loading provides a high number of stacked clay particles compared to those at 2–4 wt.%. These findings are correlated with the results provided by XRD curves.

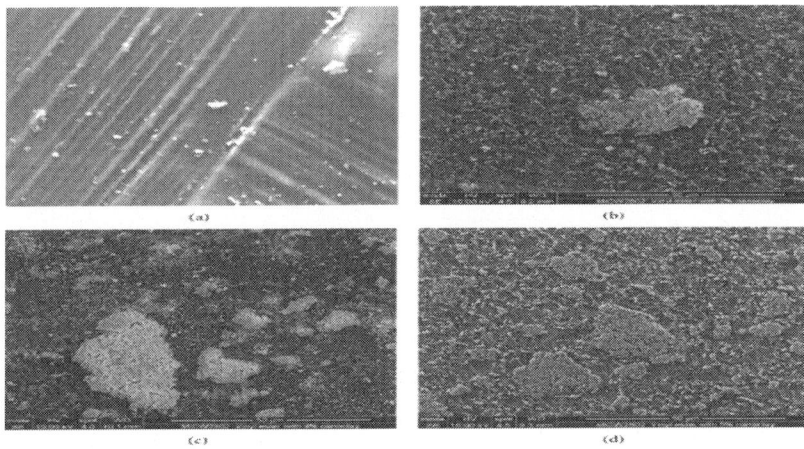

Figure 2: SEM images at 50 μm of (a) neat vinyl ester, (b) 2 wt.%, (c) 4 wt.%, and (d) 5 wt.% nanocomposites.

Energy Dispersive X-Ray Spectrometry (EDS)

Figure 3 represents the incorporation of different fractions of layered silicate into the vinyl ester matrix. It was observed that the incorporation of the clay into the polymer sea was fairly homogeneous with a small amount of agglomerative layers at higher clay loading levels. Also, it was found that the enhancement of the clay loadings led to an increase in the amount of clay agglomeration which was attributed to the viscosity of the mixture. EDS represents the layered silicate as white points which reflected the Si element. At 2 wt.% clay loading, the level of distribution of clay into the vinyl ester matrix was uniform and no agglomeration layers were observed at 55x magnification of EDS. By the presence of more clay (i.e., 4 wt.%), the nanocomposite structure exhibited a reasonable intercalation/exfoliation structure although the aggregation of a few layers was obtained. In addition, the incorporation of high amounts of clay such as 5 wt.% led to reducing the homogeneity and increasing the aggregation and the microvoids in the nanocomposite structure. These results explain the reduction in the d-spacing value as was calculated by XRD and confirmed the results by SEM.

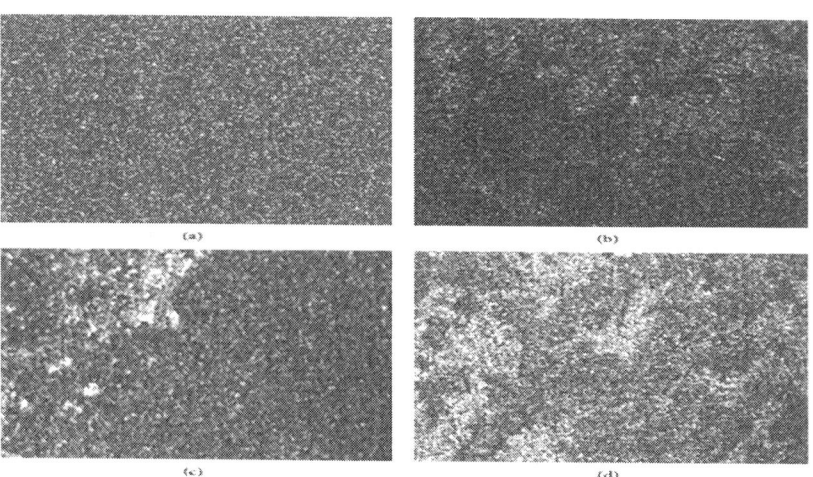

Figure 3: EDS images at 55x magnification of (a) neat vinyl ester (Cl element), (b) 2 wt.% (Si element), (c) 4 wt.% (Si element), and (d) 5 wt.% (Si element) nanocomposites.

Transmission Electron Microscopy (TEM)

Figure 4 shows the TEM micrographs of 2, 4, and 5 wt.% nanocomposite samples at 50 nm, where the bright area corresponds to the matrix sea and the dark lines signify the stacked or individual silicate layers. Indications from the higher magnification images are that greater levels of exfoliation nanocomposites are achieved with lower nanoclay loading. At 2 wt.% clay loading, the TEM image exhibits uniform distribution of layered silicate throughout the vinyl ester matrix. An intercalated/exfoliated structure is observed at 4 wt.% clay loading, as seen in Figure 4. The layered silicate shows well-dispersed clay with a few aggregation layers. At 5 wt.% clay loading, additional dark regions are observed which indicate the aggregation of silicate layers and insufficient dispersion. TEM images conclude that the particle lumps (agglomeration) are increased by the addition of more than 4 wt.% clay loading. This was attributed to the high viscosity of the mixture where the ability to mix the layered silicate and the polymer is restricted. It is acceptable that the higher the amount of clay loading mixed with the polymer, the less exfoliated and aggregated the nanocomposite structure [25, 26]. These outcomes support the results by XRD, SEM, and EDS.

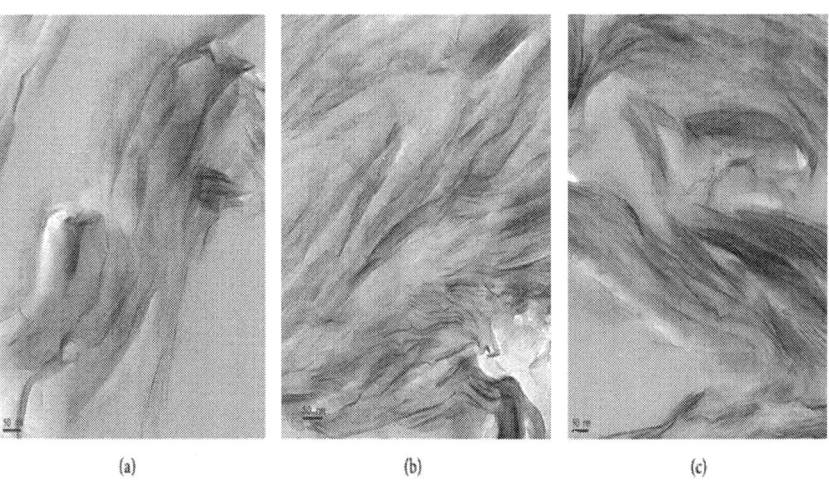

(a)   (b)   (c)

Figure 4: TEM micrographs at 50 nm magnification of (a) 2 wt.%, (b) 4 wt.%, and (c) 5 wt.% nanocomposites.

Low Velocity Impact Response

In this study, the falling weight impact tests, which are the most common for composite and nanocomposite materials, were carried out on the neat vinyl ester and the corresponding nanocomposites. Different clay concentrations were used which included 2, 4, and 5 wt.% clay loading. The impact test was used in order to analyse and evaluate the effect of the incorporation of layered silicate into the polymer matrix. Many parameters can be calculated from this test, such as the maximum force (N) and energy absorption (J) in a function of time (ms) curve. Four specimens from each group were tested and the average values were calculated as seen in Table 2. From Table 2, although the amount of load was high, the effect of the incorporation of the layered silicate into the vinyl ester matrix was observable and the addition of clay resulted in enhancing the impact properties. The maximum force and energy absorption were increased up to 42% and 59.74% respectively at 4 wt.% clay loading. The improvements of the maximum load and energy were proportional to the amount of the clay loading; however, at higher amounts of clay loading such as 5 wt.%, the force and energy absorption decreased compared to the 4 wt.% clay concentration, as seen in Figures 5 and 6. These results indicated that the addition of nanofillers to the polymer matrix not only increases the strength of the output product but also increases the toughness.

Table 2: Impact test results of neat polymer and the corresponding of nanocomposites

Samples	Fmax (N)	Improvement of the peak load (%)	Energy (J)	Improvement of the energy absorption (%)
Neat vinyl ester	1327 ± 9.00	0	1.54 ± 0.38	0
Vinyl ester + 2 wt.% clay loading	1501 ± 8.80	13.11	1.83 ± 0.31	18.9

Vinyl ester + 4 wt.% clay loading	1885 ± 9.00	42.05	2.46 ± 0.30	59.74
Vinyl ester + 5 wt.% clay loading	1655 ± 10.0	24.72	1.95 ± 0.33	21.03

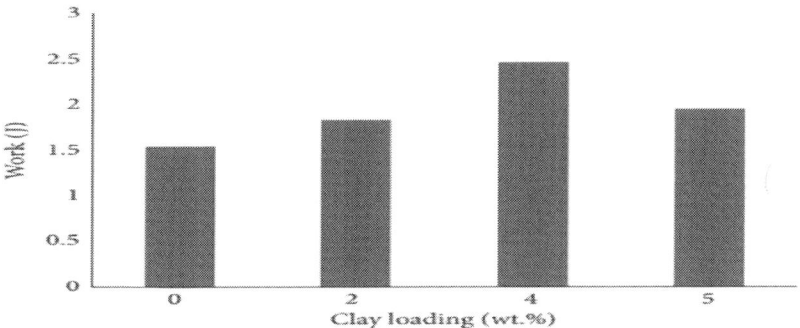

Figure 5: The relation between the work (J) and the clay loading wt.%.

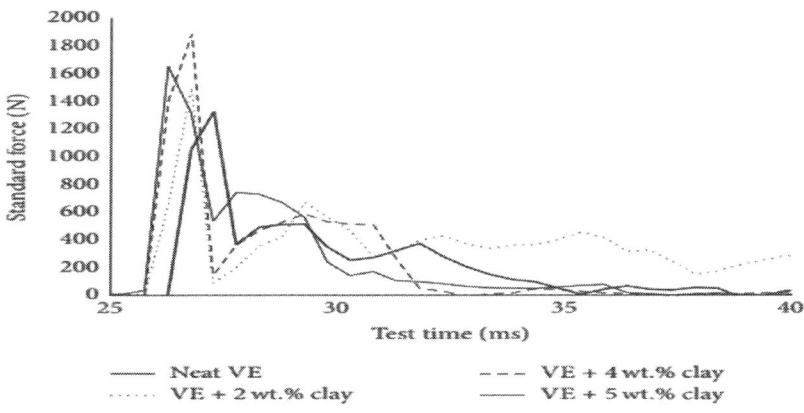

Figure 6: Maximum force versus time traces of the impact test.

One reason for the enhancement of the impact properties can be traced to the existence of microvoids while mixing the nanolayers and the polymer. When the impact load was applied, the microvoids initiated the shear yielding of the combinations of vinyl ester polymer and the layered silicate throughout the whole volume and at the start of the crack propagation. Thus, the shear yielding distributed the mechanical stress and enhanced the strength and toughness of the nanocomposites by absorbing the energy [27]. Nanocomposites voids were calculated by using the following equation:

$$V_v = \rho_n \left(\frac{w_c}{\rho_c} + \frac{w_m}{\rho_m} \right),$$

(2)

where V_v is the volume fraction of voids, ρ_n is the density of nanocomposites, w_c is the weight percent of clay (%), ρ_c is the density of clay g/cm^3 (0.16), w_m is the weight percent of matrix (%), and ρ_m is the density of matrix g/cm^3. The density of nanocomposites was calculated as follows:

$$\rho_n = \rho_c w_c + \rho_m w_m.$$

(3)

The voids content in nanocomposites felid is a big concern for the industrial or engineering designs. The pre-failure and the promoting of the local deformation of the applications can be obtained by the existence of high content of microvoids [28]. Thus the study of the parameters influences the content voids as well as the percentage of these values can help to protect the presence of such drawback.

From (2) and (3), the voids content in the nanocomposites sample was calculated and presented in Table 3. It can be seen that the voids percentage was proportional to the clay content. Maximum voids volume was found at 5 wt.% clay loading which represented 9.03%. The voids percentage can conclude that the acceptable voids content regards that the impact properties in the existence parameters used are to be less than 6.5% volume. Otherwise, less interfacial interaction between the layered silicate and the polymer will take place. Also, it may be noted that the tortuous path of the clay layers when the interface between them took place played an important role in the distribution of the mechanical stress applied. As concluded by SEM, the distance between particles and clay volume has an inverse relationship where the distance is decreased by the addition of more clay. Thus, the tortuous path is increased and the crack propagation would take a longer path. This phenomenon explains the improvement of the impact properties at 4 wt.% compared to 2 wt.%. As a result, the intercalation system can provide better impact properties than the exfoliation system [29].

Table 3: Voids content of different nanocomposites samples

Sample	Void content (%)	Percentages (%)
Vinyl ester + 2 wt.% clay	2.45	1.32
Vinyl ester + 4 wt.% clay	6.40	5.06
Vinyl ester + 5 wt.% clay	9.03	7.55

At 5 wt.% clay loading, the impact properties were reduced which was ascribed to the presence of the aggregation layers where the stress concentration factor was high. When the stress concentration factor is high, the initiation of premature failure may happen. In addition, the microvoids have contradictory functions regarding the impact properties. Fewer amounts of microvoids will allow the yield shielding of the applied load to be presented. However, the high amount of microvoids (i.e., 5 wt.%) will reduce the interfacial interaction between the polymer and layered silicate, so premature failure will dominate. Although the reduction of impact properties at high loadings of clay was observed, still there was an improvement compared to neat vinyl ester. This study was in close agreement to the report conducted by Lin et al. [30].

The results can be correlated to the SEM. Figure 2 shows the SEM images of neat matrix and the corresponding nanocomposite. It can be seen that the nanocomposite samples had microphases in their structure, whereas the neat matrix showed a smooth glassy structure, which was attributed to the brittleness of the material.

Fragmentation Characteristics

The impact fracture shapes of different samples are shown in Figure 7. As can be seen, the neat polymer exhibited brittle fractures as the samples were fragmented when undergoing the impact load. The diameter of the hole presented by the impact force for neat vinyl ester was almost 40 mm. At 2 wt.% clay loading, the nanocomposite samples showed better impact stability compared to neat polymer. The samples were not wholly fragmented upon impact force. The hole diameter of the 2 wt.% nanocomposite was less than neat polymer which presented about 30 mm. The best impact resistance was found at 4 wt.% clay loading where the samples showed only about a 25 mm hole. At 5 wt.% nanocomposites, the sample showed less stability upon the impact load compared to the 4 wt.% clay loading and exhibited about a 30–40 mm diameter hole.

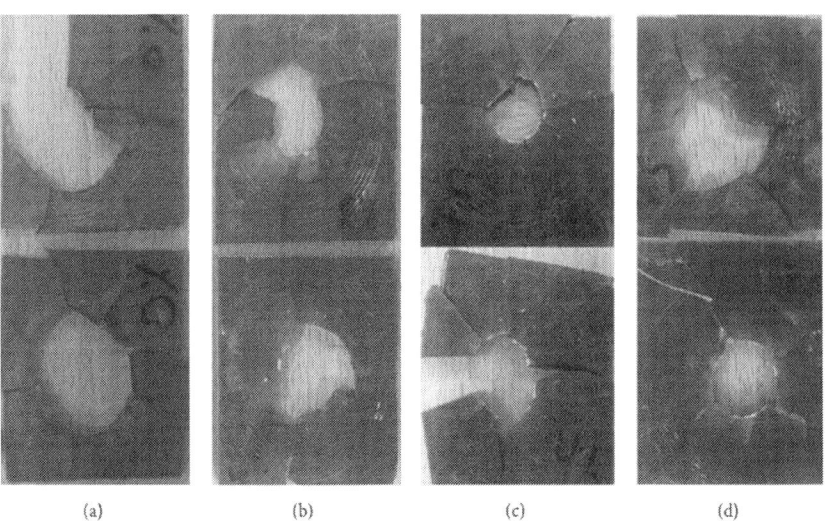

(a)   (b)   (c)   (d)

Figure 7: Fragmentation characteristics of (a) neat vinyl ester, (b) nanocomposites of 2 wt.% clay loading, (c) nanocomposites of 4 wt.% clay loading, and (d) nanocomposites of 5 wt.% clay loading.

Creep-Strain Relaxation Behaviour

The creep relaxation measurement is a key for understanding the performance of the product and the material processing. The test can help to evaluate the material's solid-like behaviour and the effect of the incorporation of layered silicate into the polymer matrix into the creep properties. In this test, the samples were subjected to a constant load and the deformation levels were calculated in a function of time. The stress relaxation processes provide an insight into the viscoelastic behaviour of the material.

Table 4 summarises the creep relaxation behaviour of neat vinyl ester and the corresponding nanocomposites at 25°C and 60°C. At 25°C, the elastic response of the nanocomposite samples exhibited less disturbance in terms of shear flow as seen in Figure 8. From the figure, it can be seen that the pristine polymer presented higher interval time and imposed stress. The initial part of the curve is termed "creep curve" and the remaining behaviour is called "relaxation level." According to the creep data, the strain reduction is proportional to the clay concentration level. The neat polymer illustrated higher strain compared to the nanocomposite samples where the strain started at 23% whereas 2 wt.% and 4 wt.% nanocomposites showed 17% and 14%, respectively. In addition, the strain of the neat polymer started to increase after 40 hours by 2%, which was attributed to the temperature. However, the nanocomposites were almost stable during the mechanical stress applied and the temperature. The enhancement of the creep properties depends on many reasons, such as the level of intercalation between the clay and polymer, the clay source, and the clay shape [31]. The microstructural changes in the clay suspension can also help to improve the creep behaviour. In addition, the presence of the layered silicate helps to improve the microphase separation, so enhancement of the elasticity took place and the reduction of the stress relaxation process was observed. Also, the layered silicate restricted the motion of the polymer chains which help to withstand the mechanical stress.

Table 4: The strain amount of the neat vinyl ester and the corresponding nanocomposites during the creep test at different interval times and temperatures

Sample	Initial strain (%)		Strain at 40 hours (%)		Strain at 60 hours (%)		Strain at 80 hours (%)		Strain at 95 hours (%)	
	25°C	60°C	25°C	60°C	25°C	60°C	25°C	60°C	25°C	60°C
Neat vinyl ester	23	17	25	21	26	22	26	22	26.3	22.5
Vinyl ester + 2 wt.% clay	17	13	17.2	13.2	17.3	13.3	17.5	13.4	18	13.6
Vinyl ester + 4 wt.% clay	14	12	14.2	12	14.4	12.1	14.6	12.2	15	12.4

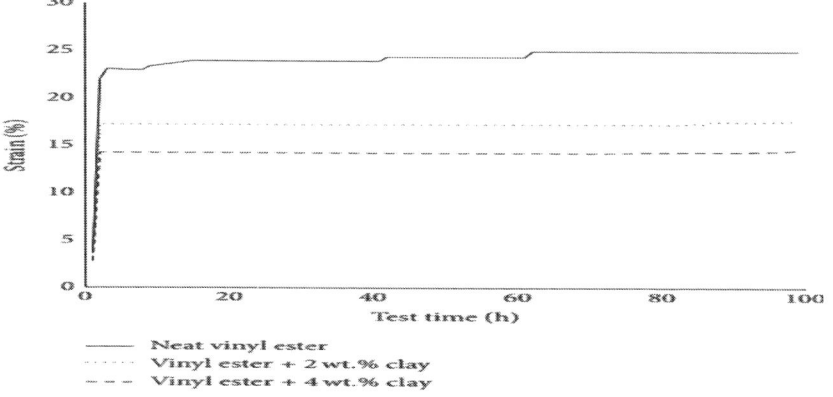

Figure 8: Creep relaxation behaviour of neat polymer and the corresponding nanocomposites at 25°C.

A similar test of the previous creep behaviour was undertaken at 60°C in order to evaluate the influence of the temperature on the neat and nanocomposite samples. Table 4 and Figure 9 show the creep relaxation behaviour of the selected samples. In the same case as the 25°C, the nanocomposites exhibited good stability under the imposed stress. The higher temperature (i.e., 60°C) represents better creep behaviour of the nanocomposite samples compared to 25°C, which may be attributed to the thermodynamic barrier where the enthalpic gain is translated into entropic gain. Thus, enhancement of the conformational links between the layered silicate and polymer took place [32]. Also, the vinyl ester may be reorienting itself into a more ordered or compact structure resulting in a stronger cross-linking plastic at higher temperature. It can be seen that the neat vinyl ester started to be deformed at 17% of strain enlargement, whereas 2 wt.% and 4 wt.% clay loading represented an initial deformation at 13% and 12%, respectively. The neat polymer showed less stability where the temperature affected the sample and led to increase the deformation by 20.5% at 40 hours. At 83 hours, the base vinyl ester started again to deform and the strain increased by 22%. However, the nanocomposites exhibited almost the same deformation level at the initial and end time.

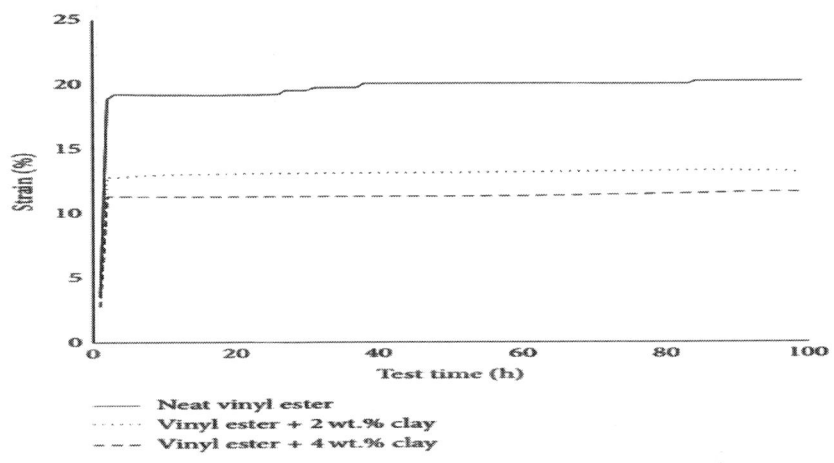

Figure 9: Creep relaxation behaviour of pristine matrix and 2 and 4 wt.% nanocomposites at 60°C.

CONCLUSIONS

The layered silicate plays an important role in terms of the polymer properties. With the addition of only small amounts of clay, the impact and creep relaxation behaviours were remarkably increased compared to the neat polymer. As far as low velocity impact strength is concerned, the introduction of layered silicate contributed to increased load-bearing capability and better energy absorption compared to the neat vinyl ester samples even with the high amount of mechanical load applied. The microvoids that existed in the nanocomposites structure had contradiction function regarding the impact properties. The first one is initiating the yield shielding upon propagation at low voids content (less than 6.5%), in turn, the crack paths will be distributed and altered by theses voids. At higher voids content, the premature failure will be obtainable due to the less interfacial interaction between the layered silicate and the polymer. The creep behaviour at both low and high temperatures, such as 25°C and 60°C, represented an improvement in the nanocomposites and the enhancement was proportional to the clay content. These improvements were due to the addition of stiff fillers, where the imposed stresses can be shared by the layered silicate and the polymer. Also, the tortuous path of the clay can distribute the transferred load and restrict the motion of the polymer chains. Thus, the strain during constant load and temperature showed better stability in the nanocomposites than the neat polymer.

REFERENCES

1. T. Yu, Y. Li, and J. Ren, "Preparation and properties of short natural fiber reinforced poly(lactic acid) composites," Transactions of Nonferrous Metals Society of China, vol. 19, no. 3, pp. s651–s655, 2009. · ·

2. A. I. Alateyah, H. N. Dhakal, and Z. Y. Zhang, "Processing, properties, and applications of polymer nanocomposites based on layer silicates: a review," Advances in Polymer Technology, vol. 32, no. 4, Article ID 21368, 2013. · ·

3. S. K. Swain and A. I. Isayev, "PA6/Clay nanocomposites by continuous sonication process," Journal of Applied Polymer Science, vol. 114, no. 4, pp. 2378–2387, 2009.

4. B. Aldousiri, H. N. Dhakal, S. Onuh, Z. Y. Zhang, N. Bennett, and M. O. W. Richardson, "Effect of layered silicate reinforcement on the structure and mechanical properties of spent polyamide-12 nanocomposites," Composites Part B: Engineering, vol. 43, no. 3, pp. 1363–1367, 2012. · ·

5. L. Mat jka, "Epoxy-silica/silsesquioxane polymer nanocomposites," in Hybrid Nanocomposites for Nanotechnology, pp. 1–84, Springer, Berlin, Germany, 2009.

6. A. I. Alateyah, H. N. Dhakal, and Z. Y. Zhang, "Water absorption behaviour, mechanical and thermal properties of vinyl ester matrix nanocomposites based on layered silicate," Polymer-Plastics Technology and Engineering, vol. 53, no. 4, pp. 327–343, 2014. · ·

7. A. B. Morgan and J. W. Gilman, "Characterization of polymer-layered silicate (clay) nanocomposites by transmission electron microscopy and X-ray diffraction: a comparative study," Journal of Applied Polymer Science, vol. 87, no. 8, pp. 1329–1338, 2003. · ·

8. H. Chen, M. Zheng, H. Sun, and Q. Jia, "Characterization and properties of sepiolite/polyurethane nanocomposites," Materials Science and Engineering A, vol. 445-446, pp. 725–730, 2007. · ·

9. M. Okamoto, Polymer/Layered Silicate Nanocomposites, Rapra Technology Limited, Shrewsbury, UK, 2003.

10. S. S. Ray and M. Okamoto, "Polymer/layered silicate nanocomposites: a review from preparation to processing," Progress in Polymer Science, vol. 28, no. 11, pp. 1539–1641, 2003. · ·

11. S. K. Samudrala and S. Bandyopadhyay, "Development of development of hybrid nanocomposites for electronic applications," in Hybrid Nanocomposites for Nanotechnology, pp. 231–287, 2009.

12. S. Lapshin, S. K. Swain, and A. I. Isayev, "Ultrasound aided extrusion process for preparation of polyolefin-clay nanocomposites," Polymer Engineering and Science, vol. 48, no. 8, pp. 1584–1591, 2008. · ·

13. S. Bathula, R. C. Anandani, A. Dhar, and A. K. Srivastava, "Microstructural features and mechanical properties of Al 5083/ SiC$_p$ metal matrix nanocomposites produced by high energy ball milling and spark plasma sintering," Materials Science and Engineering A, vol. 545, pp. 97–102, 2012. · ·

14. S. Lapshin and A. I. Isayev, "Continuous process for melt intercalation of PP-clay nanocomposites with aid of power ultrasound," Journal of Vinyl and Additive Technology, vol. 12, no. 2, pp. 78–82, 2006. · ·

15. Q. T. Nguyen and D. G. Baird, "Preparation of polymer-clay nanocomposites and their properties,"Advances in Polymer Technology, vol. 25, no. 4, pp. 270–285, 2006. · ·

16. S. K. Patra, G. Prusty, and S. K. Swain, "Ultrasound assisted synthesis of PMMA/clay nanocomposites: study of oxygen permeation and flame retardant properties," Bulletin of Materials Science, vol. 35, no. 1, pp. 27–32, 2012. · ·

17. A. I. Alateyah, H. N. Dhakal, and Z. Y. Zhang, "Mechanical and thermal properties characterisation of vinyl ester matrix nanocomposites based on layered silicate," World Academy of Science, Engineering and Technology, vol. 81, pp. 1–8, 2013.

18. F. C. Liu, E. H. Han, W. Ke et al., "Polar influence of the organic modifiers on the structure of montmorillonite in epoxy nanocomposites," Journal of Materials Science and Technology, vol. 29, no. 11, pp. 1040–1046, 2013. · ·

19. T.-T. Doan, H. Brodowsky, and E. Mäder, "Jute fibre/polypropylene composites II. Thermal, hydrothermal and dynamic mechanical behaviour," Composites Science and Technology, vol. 67, no. 13, pp. 2707–2714, 2007. · ·

20. B. K. Deka and T. K. Maji, "Study on the properties of nanocomposite based on high density polyethylene, polypropylene, polyvinyl chloride and wood," Composites Part A: Applied Science and Manufacturing, vol. 42, no. 6, pp. 686–693, 2011. · ·

21. T. D. Fornes, P. J. Yoon, H. Keskkula, and D. R. Paul, "Nylon 6 nanocomposites: the effect of matrix molecular weight," Polymer, vol. 42, no. 25, pp. 9929–9940, 2001.

22. B. Wetzel, F. Haupert, K. Friedrich, M. Q. Zhang, and M. Z. Rong, "Impact and wear resistance of polymer nanocomposites at low filler content," Polymer Engineering and Science, vol. 42, no. 9, pp. 1919–1927, 2002. · ·

23. P. Domone and J. Illston, Construction Materials: Their Nature and Behaviour, Taylor & Francis, New York, NY, USA, 2010.

24. A. Almagableh, P. R. Mantena, A. Alostaz, W. Liu, and L. T. Drzal, "Effects of bromination on the viscoelastic response of vinyl ester nanocomposites," Express Polymer Letters, vol. 3, no. 11, pp. 724–732, 2009. · ·

25. X. Liu and Q. Wu, "PP/clay nanocomposites prepared by grafting-melt intercalation," Polymer, vol. 42, no. 25, pp. 10013–10019, 2001. · ·

26. J. Jordan, K. I. Jacob, R. Tannenbaum, M. A. Sharaf, and I. Jasiuk, "Experimental trends in polymer nanocomposites—a review," Materials Science and Engineering A, vol. 393, no. 1-2, pp. 1–11, 2005. · ·

27. G. Hartwig, "Fracture behavior of polymers," in Polymer Properties at Room and Cryogenic Temperatures, pp. 187–218, Springer, New York, NY, USA, 1994.

28. P. Ladeveze, "A damage computational approach for composites: basic aspects and micromechanical relations," Computational Mechanics, vol. 17, no. 1-2, pp. 142–150, 1995. · ·

29. A. S. Zerda and A. J. Lesser, "Intercalated clay nanocomposites: morphology, mechanics, and fracture behavior," Journal of Polymer Science, Part B: Polymer Physics, vol. 39, no. 11, pp. 1137–1146, 2001. · ·

30. J. C. Lin, L. C. Chang, M. H. Nien, and H. L. Ho, "Mechanical behavior of various nanoparticle filled composites at low-velocity impact," Composite Structures, vol. 74, no. 1, pp. 30–36, 2006. · ·

31. S. T. Lim, C. H. Lee, H. J. Choi, and M. S. Jhon, "Solidlike transition of melt-intercalated biodegradable polymer/clay nanocomposites," Journal of Polymer Science, Part B: Polymer Physics, vol. 41, no. 17, pp. 2052–2061, 2003. · ·

32. M. Ganß, B. K. Satapathy, M. Thunga, R. Weidisch, P. Pötschke, and A. Janke, "Temperature dependence of creep behavior of PP-MWNT nanocomposites," Macromolecular Rapid Communications, vol. 28, no. 16, pp. 1624–1633, 2007. · ·

An Exploratory Study of Tridentate Amine Extractants: Solvent Extraction and Coordination Chemistry of Base Metals with Bis((1R-benzimidazol-2-yl)methyl)amine

Nomampondo P. Magwa[1], Eric Hosten[2], Gareth M. Watkins[1], and Zenixole R. Tshentu[1]

[1]Department of Chemistry, Rhodes University, Grahamstown, South Africa

[2]Department of Chemistry, Nelson Mandela Metropolitan University, Port Elizabeth, South Africa

ABSTRACT

Solvent extraction of base metals using bis((1-decylbenzimidazol-2-yl)methyl)amine (BDNNN) showed a lack of pH-metric separation of the metals. The extraction system was described quantitatively using the equilibria involved to derive the mathematical explanation for the two linear log D vs pH$_e$ **plots for each metal ion extraction curve, and coordination numbers could also be extracted from the two slopes. The lack of separation was attributed to the absence of stereochemical "tailor making" since the complexes isolated from the reaction of the ligand, bis((1H-benzimidazol- 2-yl)methyl)amine (NNN), with base metals suggested the formation of similar octahedral complex species from spectral and crystal structure evidence. The bis tridentate coordination observed was in agreement with information extracted from the extraction data. This investigation opens up an opportunity and an approach for the evaluation of amines as extractants but cautions against tridentate ligands.**

INTRODUCTION

There is an increasing demand for substantial production of metals of high purity in the metallurgical industries in an attempt to meet the corresponding demand for both industrial and domestic applications. This has been the catalyst for the development of simpler and more economical routes for purification of metal ions from their ore solutions. The scale for example of nickel operations is limited, and therefore one has to maximise profit via processing efficiency rather than volume. The ultimate goal is to design processes that are environmentally friendly, cost-effective, time-saving and selective. Solvent extraction, a widely applied technique for the recovery of base metals [1,2], meets some of these principles since high boiling point and high flashpoint solvents are used and are recycled together with an extractant after the stripping step, and the system can be tuned through appropriate extractant design to achieve selectivity [3]. The basis of base metal ion separation is a unique property of the coordination chemistry of the particular metal ion. In the development of a metal ion specific extractant, it is necessary to consider the characteristics of the metal ions from which the desired metal ion must be removed as well as its own [3]. Improvements of the chemical processes in the solvent extraction system require a thorough knowledge and investigation of the chemistry involved in order to achieve a meaningful advancement of this technology.

There has been an envisaged shift towards aminebased extractants, and this is motivated by the favourable properties that are offered by the nitrogenous ligands especially the aromatic amines as compared with oxygenbased extractants [3-5]. Some of these properties include the intermediate pK_a values as well as s and p bonding capabilities resulting in extractions in the low pH range and in a possibility of separation through bonding preferences respectively. The latter property has been dubbed "stereochemical tailor-making" by du Preez [5]. On the other hand, the strong aliphatic amine ligands with - donor only character show lack of relative preference for the metal ions but this can be improved if chelates are used, and also tend to form metal complexes at relatively high pH values which is undesirable [5]. The chelate effect has also been exploited effectively in a bidentate aromatic system (1-octyl-2,2'-pyridylimidazole) providing for effective separation of nickel from other base metals in strong acidic sulfate solutions with a possibility of back-extraction [3].

An extension of these systems to tridentate ligands would have an additional advantage of increasing the extraction equilibrium constants [6] due to the high complex formation constants for reactions of base metals and tridentate ligands [7,8], thereby requiring relatively low extractant-to-metal ratios to achieve quantitative extractions [6]. The only example of a tridentate amine extractant in the literature is that of a derivative of diethylenetriamine [9] but extractions occurred at relatively high pH values as expected due to high pK_a values of aliphatic amines [8], and the small $pH_{0.5}$ values implied a lack of pH-metric separation of the later 3d metal ions [9]. It would be hoped that good separation factors of the metal chelates would be achieved through stereochemical considerations since nickel(II) is known to form the most stable spin free octahedral (O_h) complexes of all base metal ions [10] while the copper(II) and cobalt(II) ions tend to form stable tetrahedral (T_d) complexes [4,11]. This study, therefore, also interrogates the coordination chemistry aspects of base metals, that infuence the extractions, with tridentate amine-based ligands.

In this account, we present the extractive and coordination chemistry of an aromatic tridentate ligand, bis((1H-benzimidazol-2-yl)methyl) amine (NNN) (Figure 1(a)), towards base metals in a sulfate/sulfonate medium. Dinonylnaphthalene sulfonic acid (DNNSA) (Figure 1(b)) was used as a bulky anion to ion-pair and transfer the cationic complexes formed in this extraction system (ion-association system) to the organic phase since the sulfate ion is known to have high hydration energies leading to lack of phase transferability [12]. The sulfate medium has become particularly important to explore since it is encountered in liquors produced in sulfatebased high nickel matte leach processes or those produced in sulfuric acid pressure leaching of laterites [13].

EXPERIMENTAL

Materials

o-Phenylenediamine (99.5%, Sigma-Aldrich), iminodiacetic acid (98%, Sigma-Aldrich), hydrochloric acid

(32%, Merck Chemicals), ammonia (28%, Merck Chemicals), methanol (99%, Sigma-Aldrich), and ethyl acetate (98%, Sigma-Aldrich) were reagent grade chemicals used as received for the synthesis of bis((1H-benzimidazol- 2-yl)methyl)amine (NNN). The reagent grade octylbromide (98%) was also obtained from Sigma-Aldrich and used as received for the synthesis of the extractant. $NiSO_4 \cdot 6H_2O$ (98%), $CuSO_4 \cdot 5H_2O$ (99%) were obtained from Merck chemicals. $CoSO_4 \cdot 7H_2O$ (97.5%) was obtained from Fluka, while $ZnSO_4 \cdot 7H_2O$ (99.5%) was obtained from BDH Chemicals. The copper(II), nickel(II), cobalt(II) and zinc(II) perchlorate hexahydrate salts which were used to prepare the metal(II) sulfonate salts using toluene-4-sulfonic acid (98%, Sigma-Aldrich) were obtained from Sigma-Aldrich. Analytical grade reagents were used without further purification in the preparations of the 0.10 M metal ion stock solutions in 3 M H_2SO_4 solution. The ICP/AAS 1000 ppm metal standards, dissolved in 0.5 N nitric acid, were used to prepare standard solutions for the construction of calibration curves using distilled, deionized, milliQ water for the dilutions. Dinonylnaphthalene sulfonic acid (50 wt% in heptane), Shellsol 2325 (17 - 22 v/v% aromatic content) and 2-octanol (98%) were obtained from Sigma-Aldrich, Shell Chemicals (SA) and Merck Chemicals, respectively.

Instrumentation

^{1}H NMR spectrometry was carried out on a Bruker AMX 400 MHz NMR spectrometer and reported relative to tetramethylsilane (0.00). A Vario Elementary ELIII Microcube CHNS analyser was used for elemental analyses. A Thermo Electron (iCAP 6000 Series) inductively coupled plasma (ICP) spectrometer equipped with an OES detector was used for metal ion analysis. The Labcon micro-processor controlled orbital platform shaker model SPO-MP 15 was used for contacting the two phases of extraction. The pH measurements were performed on a Metrohm 827 pH meter using a combination electrode with 3 M KCl as electrolyte. The metalcomplexes were characterised using infrared spectrome try on both Perkin Elmer 400 FTIR and 100 FTIR-ATR spectrometers. The solid reflectance spectra of the metal complexes were recorded on a Shimadzu UV-VIS-NIR Spectrophotometer UV-3100 with a MPCF-3100 sample compartment with samples mounted between two quartz discs which fit into a sample holder coated with barium sulfate. The spectra were recorded over the wavelength range of 2000 - 250 nm, and the scans were conducted at a medium speed using a 20 nm slit width. The Gallenkamp melting point apparatus (temperature range, $0°C - 350°C$) was used to measure the melting points. The conductivity measurements were carried out on a A.W.R. Smith Process Instrumentation cc Laboratory Bench Meter Model AZ 86555 with ABS graphite cell probe using an aqueous standard which has a conductivity value of 135 $ohm^{-1}·cm^2·mole^{-1}$ at $20°C$ for the calibration. All the complexes were prepared in DMF as a solvent to a concentration of 10^{-3} M for the conductivity measurements.

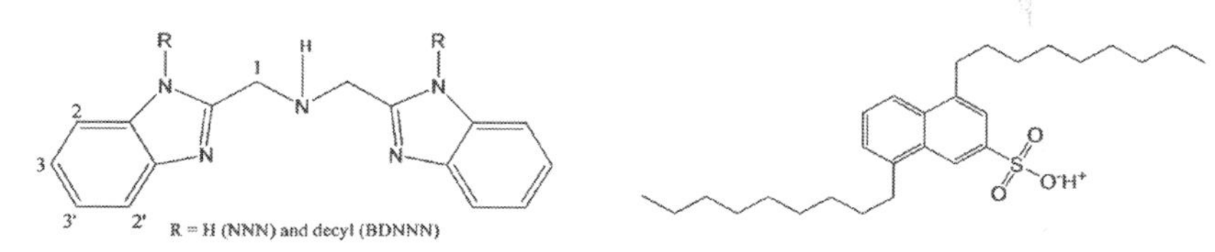

Figure 1: The chemical structures of (a) bis((1R-benzimidazol-2-yl)methyl) amine (NNN for R = H, and BDNNN for R = Decyl) and (b) dinonylnaphthalene sulfonic acid (DNNSA).

Synthesis of the Ligand and Extractant

Bis ((1H-Benzimidazol-2-yl)Methyl)Amine (NNN)

The bis((1H-benzimidazol-2-yl)methyl)amine (NNN) was synthesized as reported elsewhere [14], except that iminodiacetic acid was used and the decolorization step using activated charcoal in methanol was necessary. The characterization data for the white precipitate of the free base was as follows: Yield: 84%, m.p., 268°C - 270°C. Anal. Calcd. for $C_{16}H_{16}N_4OS$ (%): C, 65.10; H, 5.80; N, 23.70. Found: C, 65.78; H, 5.80; N, 23.07. $^1H \cdot NMR$ (CDCl$_3$) d (ppm): 4.06 (4H, s, H1), 7.15 (4H, m, H3,H3'), 7.52 (4H, m, H2,H2'). IR (cm^{-1}): 3208 v(N-H), 3049 v(sec N-H), 1592 v(C = N).

Bis ((1-Decylbenzimidazol-2-yl)Methyl)Amine (BDNNN)

The alkylated derivative of the ligand was prepared according to a literature method [15]. However, the purifycation step was carried out as follows: The resulting solution after the removal of the KBr salt was concentrated via rotary evaporation, and purified using a silica gel chromatographic column with ethyl acetate/methanol (4:1) solvent system. After the removal of the solvent by rotary evaporation the product was obtained as brown oil. Yield = 67%. Anal. Calcd. for $C_{36}H_{59}N_5O_2$ (%): C, 72.81; H, 10.01; N, 11.79. Found: C, 72.95; H, 10.61; N, 11.86. ^1H NMR (CDCl$_3$) δ (ppm): δ 0.89 (6H, t, CH$_3$), 1.13 (24H, m, CH$_3$(CH$_2$)$_6$), 1.01(4H, t, CH$_2$-CH$_3$), 1.55(4H, t, CH$_2$-CH$_2$N), 3.96 (4H, t, CH$_2$-N), 4.02 (4H, s, H1), 7.26 (4H, q, H3 & H3'), 7.28 (2H, d, H2), 7.75 (2H, d, H2'). IR (cm^{-1}): 1522 v(C = N).

Syntheses of Metal Complexes

All the reactions for the formation of coordination complexes (sulfonate and sulfate compounds) were conducted in absolute ethanol using the toluene-4-sulfonate salts of the metals, and inert conditions were adopted for the synthesis. 10 mL of hot ethanolic solution (60°C) containing 2 mmol of the ligand was added dropwise to 10 mL of the metal ion solution (1 mmol, respectively for each metal ion). The mixture was heated under reflux overnight and the precipitate that formed was filtered off, washed with ethanol and dried.

Sulfate Complexes

$[Co(NNN)_2]SO_4 \cdot 4H_2O$: Color: pink. Yield = 51%, m.p., 252°C - 254°C. Anal. Calcd. for $C_{32}H_{38}N_{10}CoO_8S$ (%):C, 49.17; H, 4.90; N, 17.92; S, 4.10. Found: C, 49.20; H, 4.46; N, 17.85; S, 4.86. IR (cm^{-1}): 3242 v(N-H), 1545 v(C = N), 1037 v_3(SO$_4$), 226 v(M-N). Conductivity (10^{-3} M, ohm^{-1}·cm^2·mole^{-1}): 66.

$[Ni(NNN)_2]SO_4 \cdot 3H_2O$: Color: purple. Yield = 57%, m.p., 253°C - 255°C. Anal. Calcd. for $C_{32}H_{36}N_{10}NiO_7S$ (%): C, 50.34; H, 4.75; N, 18.35; S, 4.20. Found: C, 50.64; H, 4.74; N, 18.28; S, 4.10. IR (cm^{-1}): 3234 v(N-H), 1538 v(C=N), 1037 v_3(SO$_4$), 230 v(M-N). Conductivity (10^{-3} M, ohm^{-1}·cm^2·mole^{-1}): 69.

$[Cu(NNN)_2]SO_4 \cdot 7H_2O$: Color: green. Yield = 65%, m.p., 228°C - 229°C. Anal. Calcd. for $C_{32}H_{44}N_{10}CuO_{11}S$ (%): C, 45.74; H, 5.28; N, 16.67; S, 3.82. Found: C, 45.81; H, 5.11; N, 16.41; S, 3.44. IR (cm^{-1}): 3233 v(N-H), 1565 v(C = N), 1060 - 1088 v_3(SO$_4$), 224 v(M-N). Conductivity (10^{-3} M, ohm^{-1}·cm^2·mole^{-1}): 71.

$[Zn(NNN)_2]SO_4 \cdot 11H_2O$: Color: white. Yield = 64%, m.p., 221°C - 222°C. Anal.Calcd. for $C_{32}H_{52}N_{10}ZnO_{15}S$ (%): C, 42.04; H, 5.73; N, 15.31; S, 3.51. Found: C, 41.96; H, 5.71; N, 15.00; S, 3.59. IR (cm^{-1}): 3227 v(N-H), 1548 v(C = N), 1037 v_3(SO$_4$), 222 v(M-N). Conductivity (10^{-3}M, ohm^{-1}·cm^2·mole^{-1}): 82.

Sulfonate Complexes

$[Co(NNN)_2](RSO_3)_2 \cdot 4H_2O \cdot 2EtOH$: Color: red. Yield = 71%, m.p., 225°C - 226°C. Anal. Calcd. for $C_{50}H_{64}N_{10}$- CoO$_{12}$S$_2$ (%): C, 53.66; H, 5.67; N, 12.52; S, 5.73. Found: C, 53.60; H, 5.34; N, 12.67; S, 5.14. IR (cm^{-1}): 3311 v(N-H), 1551 v(C = N), 1150-1161 v_3(RSO$_3$), 279 v(M-N). Conductivity (10^{-3} M, ohm^{-1}·cm^2·mole^{-1}): 136.

$[Ni(NNN)_2](RSO_3)_2 \cdot 3H_2O \cdot 2EtOH$: Color: purple. Yield = 58%, m.p., 246°C - 248°C. Anal. Calcd. for $C_{50}H_{62}$- N$_{10}$NiO$_{11}$S$_2$ (%): C, 54.50; H, 5.67; N, 12.71; S, 5.82. Found: C, 54.43; H, 5.58; N, 12.47; S, 5.70. IR (cm^{-1}): 3313 v(N-H), 1550 v(C = N), 1151 - 1164 v_3(RSO$_3$), 246

v(M-N). Conductivity (10^{-3} M, ohm^{-1}·cm^2·mole^{-1}): 139.

[Cu(NNN)$_2$](RSO$_3$)$_2$·12H$_2$O: Color: blue. Yield = 68%, m.p., 201°C - 202°C. Anal. Calc. for C$_{46}$H$_{68}$N$_{10}$CuO$_{18}$S$_2$ (%): C, 46.95; H, 5.82 N, 11.90; S, 5.45. Found: C, 46.48; H, 5.47; N, 11.76; S, 5.85. IR (cm^{-1}): 3213 v(N-H), 1550 v(C = N), 1147 - 1172 v_3(RSO$_3$), 262 v(M-N). Conductivity (10^{-3} M, ohm^{-1}·cm^2·mole^{-1}): 141.

[Zn(NNN)$_2$](RSO$_3$)$_2$·3H$_2$O·2EtOH: Color: white. Yield = 62%, m.p., 201°C - 202°C. Anal. Calc. for C$_{50}$H$_{62}$N$_{10}$- ZnO$_{11}$S$_2$ (%): C, 54.17; H, 5.64; N, 12.63; S, 5.78. Found: C, 54.50; H, 5.64; N, 12.30; S, 5.53. IR (cm^{-1}): 3304 v(N-H), 1549 v(C=N), 1172 - 1187 v_3(RSO$_3$), 258 v(M-N). Conductivity (10^{-3} M, ohm^{-1}·cm^2·mole^{-1}): 147.

Solvent Extraction Procedure

All the extractions were carried out at 25°C (±1°C) in a temperature controlled laboratory. Equal volumes (10 mL) of 0.001 M metal ion solution (aqueous layer) and 80% 2-octanol/shellsol solution (organic layer containing the extractant BDNNN and the counterion DNNSA) were pipetted into 50 mL conical separating funnels. The contents in the funnel were shaken using an automated orbital platform shaker for 30 minutes at an optimised speed of 200 rpm. A minimum period of 60 minutes was observed before harvesting the raffinates. The raffinates were filtered through a 33 mm millex-HV Millipore filter (0.45 µm) and diluted appropriately for analysis by ICP. The percentage extractions (%E) of the metal ions were calculated from the concentrations of the metal ions in the aqueous phase using equation 1 below:

$$\%E = \left(\frac{C_i - C_s}{C_i} \right) \times 100$$

(1)

where C_i is the initial solution concentration (mg/L) and C_s is the solution concentration after extraction.

The extraction efficiencies were investigated as a function of pH, and all the extraction curves were plotted with Sigma Plot 11.0.

X-Ray Structure Determination and Refinement

Single crystals of $[Cu(NNN)_2](RSO_3)_2 \cdot 12H_2O$, suitable for X-ray diffraction, were obtained by slow evaporation of the ethanolic mother liquor of this complex at room temperature. X-ray diffraction studies were performed at 200 K using a Bruker Kappa Apex II diffractometer with graphite monochromated Mo Kα radiation (λ = 0.71073 Å). The crystal structures were solved by direct methods using SHELXTL [16]. All non-hydrogen atoms were refined anisotropically. C-bound H atoms were placed in calculated positions and refined as riding atoms, with bond lengths 0.95 (aromatic CH), 0.99 (CH_2), 0.98 (CH_3) Å and with U_{iso}(H) = 1.2 (1.5 for methyl) U_{eq}(C). Hydrogens bonded to nitrogen were located on a Fourier map and allowed to refine freely. Hydrogens on water molecules were restrained to an O-H bond length of 0.84 Å and H-O-H angle of 110°. Diagrams and publication material were generated using SHELXTL, PLATON [17] and ORTEP-3 [18].

RESULTS AND DISCUSSION

Solvent Extraction Studies

These studies were carried out in dilute synthetic sulfate solutions. The extractant (BDNNN) was used for the extraction studies while the ligand (NNN) was used to study the coordination chemistry involved. The complexes of the extractant were oily and not easily isolated hence we used the NNN ligand for the coordination chemistry studies. The decyl groups of the extractant (BDNNN) would be positioned away from the coordination sphere and therefore the use of the ligand (NNN) would not change the coordination chemistry of the extractant from a steric hindrance point of view.

The use of low pK_a groups (benzimidazoles) on the ligand was expected to result in metal extractions from highly acidic solutions in comparison with tridentate aliphatic amine extractants based on diethylenetriamine [9]. This was well exploited but also resulted in lack of room for back-extraction of some metals (like Cu(II), Co(II) and possibly Ni(II)) at the "left legs" of their extraction curves. The lack of pH-metric separation of the base metal ions, however, was evident from the small $pH_{0.5}$ values (Figure 2). Interestingly, the order of the extraction of the metal ions somewhat followed the IrvingWilliams stability order [19] in the shifting of the curves towards the acidic region with the exception of Co(II) and Ni(II) extractions, which could be influenced by kinetic effects [20]. It is also noteworthy to report on the hard ions (Mn^{2+} and Mg^{2+}) extracting ability of this tridentate ligand but with rejection of Fe(III) in the pH range 0 - 2.6, which is attractive for the latter but not the former.

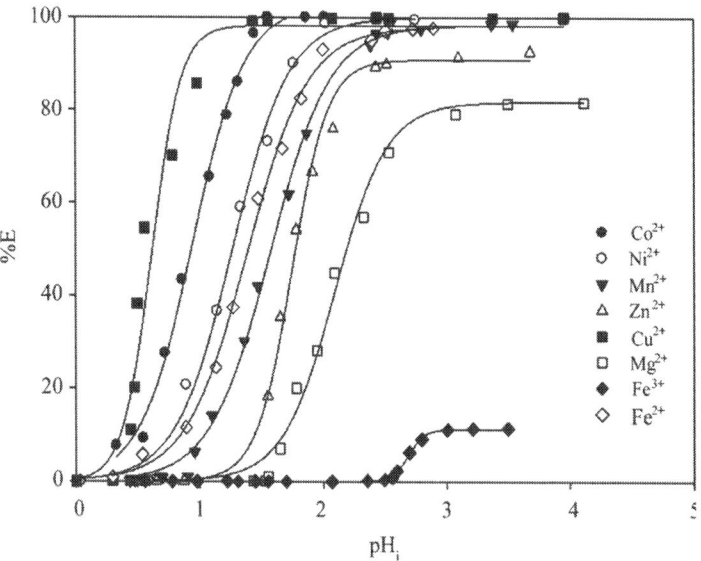

Figure 2: A plot of %E vs initial pH of equimolar concentrations (0.001 M) of Mg^{2+}, Mn^{2+}, Fe^{3+}, Fe^{2+}, Co^{2+}, Ni^{2+}, Cu^{2+} and Zn^{2+}, extracted with BDNNN (at M:L ratio 1:40) and 0.02 M DNNSA in 2-octanol/Shellsol 2325 (8:2) from a dilute sulfate medium.

It is clear, therefore, that the move towards tridentate ligands, even with low pK_a aromatic nitrogen groups and a strong aliphatic amine group, is not sufficient to successfully exploit the low pH range interactions with metal ions since extraction isotherms are pushed further into the low pH range thereby compromising back-extraction. These ligands also lack selectivity for the important borderline metal ions which can possibly be tuned through bonding preferences. The coordination chemistry studies (Section 3.2) were, therefore, conducted to try and elucidate the underlying aspects of bonding that influenced the extraction isotherms observed.

In a quantitative treatment for this solvent extraction system, similar to that applied for a chelating system (HL) [6], the protonation, complexation and phase distribution equilibria can be used to describe the system mathematically with respect to the distribution ratio of a metal ion (M^{n+}), and also give insight into the coordination numbers involved in the extraction reaction. The protonation equilibria which were studied using potentiometry in the pH range of 2 - 10 by Hay et al. [7] showed two constants for bis((1H-benzimidazol-2-yl)methyl) amine (NNN, L), and they were 5.64 and 10.12 respectively for the cumulative protonation steps (LH^+ and LH_2^{2+}). The species distribution plot, constructed from the above constants using the computer program HYSS [21], is given in Figure 3 for the pH-metric speciation involved for L in the aqueous phase.

The chelating agent (L) must distribute between the organic and aqueous phases to effect coordination in the aqueous phase, and that distribution coefficient is represented by $K_D(L)$:

$$\left(L \right)_a \ \square \ \left(L \right)_o$$

and

$$K_D \left(L \right) = \frac{\left[L \right]_o}{\left[L \right]_a} \tag{2}$$

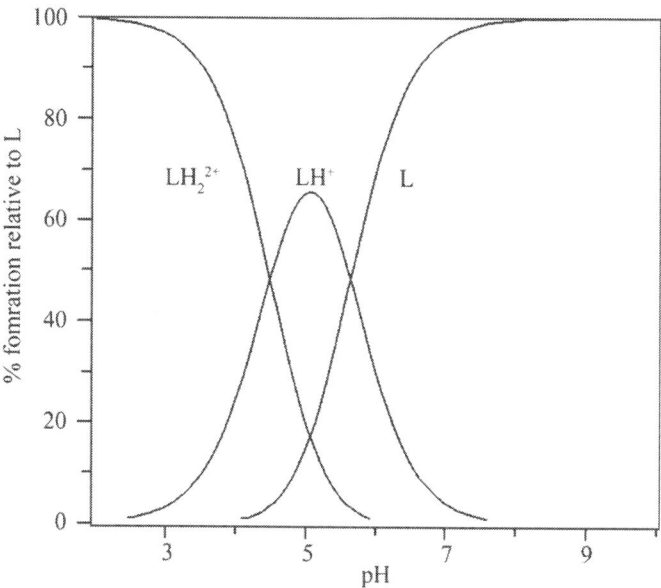

Figure 3: Protonation species distribution diagram for bis ((1Hbenzimidazol-2-yl) methyl)amine (NNN, L).

However, in the aqueous phase the following two protonation equilibria may exist depending on pH:

$$LH_2^{2+} \; \square \; H^+ + LH^+, \; K_{a1} = \frac{\left[H^+\right]_a \left[LH^+\right]_a}{\left[LH_2^{2+}\right]_a} \tag{3}$$

and

$$LH^+ \; \square \; H^+ + L \; \text{ and } \; K_{a2} = \frac{\left[H^+\right]_a \left[LH^+\right]_a}{\left[LH^+\right]_a} \tag{4}$$

Then, the metal ion chelates with the neutral ligand to form a cationic complex:

$$M^{n+} + mL \; \square \; ML_m^{n+}$$

and

$$K_f = \frac{\left[ML_m^{n+}\right]_a}{\left[M^{n+}\right]\left[L\right]_a^m}$$

(5)

It must be borne in mind, however, that the metal ion will replace proton(s) in the pH ranges under investigation but the protonation equilibria will accommodate this in the mathematical treatment. Finally, the chelate which is ion-paired by an anion (in our case two sulfonate anions represented by X^{n-}) to form an extractible species, $[ML_m]X$, distributes itself between the organic and aqueous phases:

$$\left(ML_m^{n+}\right)_a + \left(X^{n-}\right)_{o/a} \quad \Box \quad \left(ML_m X\right)_o$$

and

$$K_D\left(ML_m^{n+}\right) = \frac{\left[ML_m X\right]_o}{\left[ML_m^{n+}\right]_a}$$

(6)

The distribution ratio (D), defined as the ratio of the concentration of the total metal species in the organic phase to that in the aqueous (regardless of its mode), is given by expression 7, on the assumption that the metal chelate distributes largely in the organic phase and that the metal ion does not hydrolyse in the aqueous phase.

$$D \approx \frac{\left[ML_m X\right]_o}{\left(M^{n+}\right)_a}$$

(7)

Substituting Equations (6) and (5) respectively into Equation (7) yields Equation (8), depicting the formation constant and the concentration of the ligand in the aqueous phase as important parameters as well as the distribution coefficient of the chelate:

$$D = K_D\left(ML_m^{n+}\right) K_f \left[L\right]_a^m$$

(8)

which can be transformed to Equation (9) if Equation (2) is substituted into Equation (8), indicating that the concentration of L in the aqueous phase is dependent on its concentration in the organic phase and that its distribution between the two phases affects the distribution ratio of the complex formed:

$$D = \frac{K_D\left(ML_m^{n+}\right)K_f}{K_D\left(L\right)^m}\left[L\right]_o^m$$

(9)

However, since the extractions are carried out at low pH, it is necessary to consider the two protonation equilibria respectively because these species occur over wide pH ranges (Figure 3), and competition of metal ions with protons for the ligand occurs early with pH due to the higher formation constants [7] compared with protonation constants thereby resulting in release of the protons from the ligand. Now, substituting Equation (4), and Equations (3) and (4), respectively, into Equation (8) yields the following respective Equations (10) and (11):

$$D = K_D\left(ML_m^{n+}\right)K_f K_{a2}^m \frac{\left[LH^+\right]^m}{\left[H^+\right]_a^m}$$

(10)

And

$$D = K_D\left(ML_m^{n+}\right)K_f K_{a2}^m K_{a1}^m \frac{\left[LH_2^{2+}\right]_a^m}{\left[H^+\right]_a^{2m}}$$

(11)

Therefore, in the pH range where only one protonation equilibrium (Equation (4)) is involved then a plot of log D vs pH (from taking the logarithms of both sides in equation 10) should yield a straight line with slope m (number of ligands bonded to the metal ion M^{n+}). But in the highly acidic region where the second proton equilibrium (Equation (3)) is also active, then a plot of log D vs pH should yield a straight line with slope (2 m).

A plot of log D vs pHe (since the extraction isotherms relate to the equilibrium condition) for this extraction system is presented in Figure 4. There is a clear change in the slope of each curve and respectively become steeper with an increase in the order of stabilization of metal ions [19]. The higher pH range for Mg^{2+}, Zn^{2+}, Mn^{2+}, Fe^{2+} and Ni^{2+} is represented by a slope in the range 1 - 2 with nickel at m = 1.97 while it peaks close to 3 - 4 ($\approx$2 m) respectively at the lower pH range with the exception of nickel (slope = 15) which gets seriously affected by the extremely acidic medium. This observation is somewhat in agreement with the mathematical model described here for this complex equilibria system suggesting two ligands per metal ion are involved in the coordination. These observations (as well as those discussed below) are also in line with the protonated ligand species observed in Figure 3 (for the protonation equilibria). The extremely acidic region (affecting the copper and cobalt curves, and to a minor extent the nickel curve in the acidic end) is only characterised by a slope of ca. (3 m - 2 m) for copper and cobalt at the higher pH end (around the pH where other metal ions also have a slope close to 4). This occurred mainly because our mathematical modelling does not take into account the third protonation equilibrium which is possible under those highly acidic conditions where cobalt and copper are extracted. This third protonation constant was also not determined by potentiometry by Hay et al. [7] due to the inaccessibility of the measurement in that pH range by potentiometry. According to our mathematical treatment, involving the third equilibrium should result in a slope 3 m = 6 at the lower pH end for cobalt and copper (and to some extent nickel), however, it is much steeper with slopes of 12, 15 and 26 for cobalt, nickel and copper respectively.

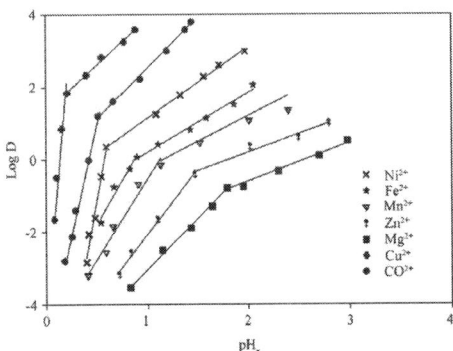

Figure 4: A plot of log D vs equilibrium pH (pH_e) for the extraction of 0.001 M M^{2+} (M = Mg^{2+}, Zn^{2+}, Mn^{2+}, Fe^{2+}, Ni^{2+}, Co^{2+} and Cu^{2+}) with 0.002 M BDNNN and 0.02 M DNNSA from sulfate medium.

It is possible, however, that not only the double (slope = 2 m) and the triple (slope = 3 m) protonation equilibria dominate in the pH range where extractions of copper and cobalt ions occur but complex multiple protonations (i m, i = 1, 2, 3, ⋯) with i exceeding 3 from the hydrogen bonding with the rings' ϖ electrons [22], hence the coordination number (m) cannot be calculated accurately from data in that pH range. It seems though that a linear plot of the points (calculated at the intersections of the two lines from each metal ion extraction data) at the vertices of the two linear plots, respectively for each metal ion, gives a negative slope ≈ 1.6 which is in agreement with the coordination numbers of 2 involved for all the metals (discussed in Section 3.2), but a mathematical treatment has not been provided herein. This observation may be coincidental, and therefore one cannot conclude on an isolated study. Once it is verified for other extraction studies with extractants similar to the one studied here, a mathematical function may be derived.

Coordination Chemistry Studies

The elemental analyses data suggested the following empirical formulae; $[M(NNN)_2]X \cdot xH_2O \cdot yEtOH$ (M = Co, Ni, Cu and Zn; X = SO_4^{2-} or $(RSO_3^-)_2$, x = 3 - 12 and y = 0 - 2). The involvement of two ligands per metal ion is in agreement with what was observed in the extraction studies. Complexes of Fe (II), Fe(III) and other hard ions could not be isolated with a good level of purity but bis-coordination is implied by the slopes of the extraction data. The molar conductivity data in DMF showed that the sulfate and sulfonate complexes have molar conductance values of 66 - 82 and 136 - 147 $\Omega^{-1} \cdot cm^2 \cdot mol^{-1}$, respectively. This indicated that all the sulfate complexes behaved as 1:1 electrolytes while the sulfonate complexes were 1:2 electrolytes [23]. This behavior in solution suggested the non-coordinated nature of the counteranions. Both the sulfate and sulfonate complexes were prepared in order to elucidate the nature of these anions respectively with respect to their innocence to coordination, and the full structural analyses are discussed in Sections 3.2.1 and 3.2.2 below. Both the sulfate and sulfonate complexes were prepared since the extraction studies were carried out in a sulfate medium and a bulky sulfonate anion was also used to replace the sulfate ion.

Spectral Analysis

The C=N stretching vibration of the benzimidazole rings of the free ligand (NNN) appeared at 1592 cm^{-1} [24,25], and coordination-induced frequencies were observed in the range 1548 - 1565 cm^{-1} upon complex formation. The lowering in the double bond character of C=N is perhaps due to the influence of the benzimidazole groups being trans to each other. The far infrared spectra of the sulfate and sulfonate complexes displayed bands in the range 222 - 279 cm^{-1} which were assigned to the ν(M-N) [26]. A strong and broad peak in the range 1137 - 1188 cm^{-1} and 1050 - 1090 cm^{-1} was present in the spectra of sulfate and sulfonate complexes respectively, and this is typical of the uncoordinated sulfate and sulfonate ions [27]. The infrared spectra of these complexes suggested that all the three donor atoms of the ligand are involved in the coordination sphere and that both the sulfate and sulfonate anions are non-coordinating. The geometry of the complexes was confirmed by UV-Vis solid reflectance electronic studies as well as by single crystal X-ray crystallography (Section 3.2.2).

Three d-d transitions that are expected in the visible region of the spectrum for an octahedral Co(II) complex are; $^4T_{1g}(F) \rightarrow {}^4T_{2g}(F)$ (v_1), $^4T_{1g}(F) \rightarrow {}^4A_{2g}(F)$ (v_2) and $^4T_{1g}(F) \rightarrow {}^4T_{1g}(P)$ (v_3) [28]. These absorption bands were observed at 1095, 544 and 482 nm respectively (Figure 5). For the nickel complexes the bands were observed at 919, 585 and 555 nm which may be assigned to $^3A_{2g}(F) \rightarrow {}^3T_{2g}(F)$ (v_1), $^3A_{1g}(F) \rightarrow {}^3T_{2g}(F)$ (v_2) and ($^3A_{2g}(F) \rightarrow {}^3T_{1g}(P)$ (v_3) transitions respectively for an octahedral symmetry [28]. The electronic spectrum of the Cu(II) complex showed one broad band at 619 nm which was ascribed to the $^2B_{1g} \rightarrow {}^2B_{2g}$ and, assuming that the second transition ($^2B_{1g} \rightarrow {}^2A_{1g}$) is masked by the intraligand transition, this is consistent with a distorted octahedral geometry [28].

X-Ray Structural Analysis

An ORTEP diagram of the crystal structure of [Cu(NNN)$_2$](RSO$_3$)$_2$·12H$_2$O is presented in Figure 6. The selected crystallographic data is presented in Table 1, and selected bond lengths and angles in Table 2.

The crystal structure of [Cu(NNN)$_2$](RSO$_3$)$_2$·12H$_2$O conclusively depicted that the complex is cationic with relatively isolated sulfonate anions (Figure 6). The closest contact that the sulfonate ion has with the cationic molecule is H(23)-O(31) = 2.560(1). The two ligands are tridentately coordinated to the copper(II) ion in the formation of a tetragonally distorted O_h geometry with the four benzimidazoles in the square plane while the two aliphatic amines occupy the apical positions. The fivemembered chelate rings have bite angles in the range 74.8° - 77.0°. The aliphatic amines are bonded trans to each other and the Cu-N lengths of 2.509 and 2.547 Å are rather long possibly due to Jahn-Teller distortion. However, the Cu-N (benzimidazole) lengths are rather short (average of 2.014 Å) due to the ϖ-acidity character of the benzimidazole group.

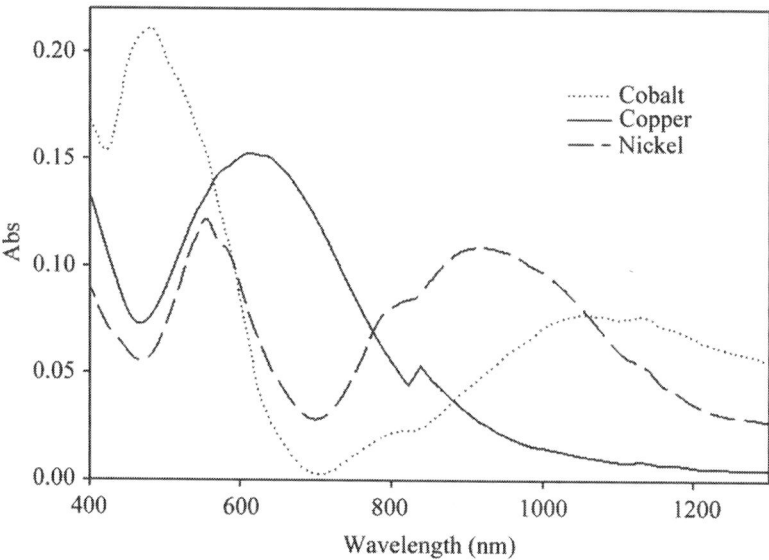

Figure 5: The UV-Vis solid reflectance spectra of $[M(NNN)_2]SO_4 \cdot xH_2O$ (M=Co, Ni and Cu; x = 3 - 12).

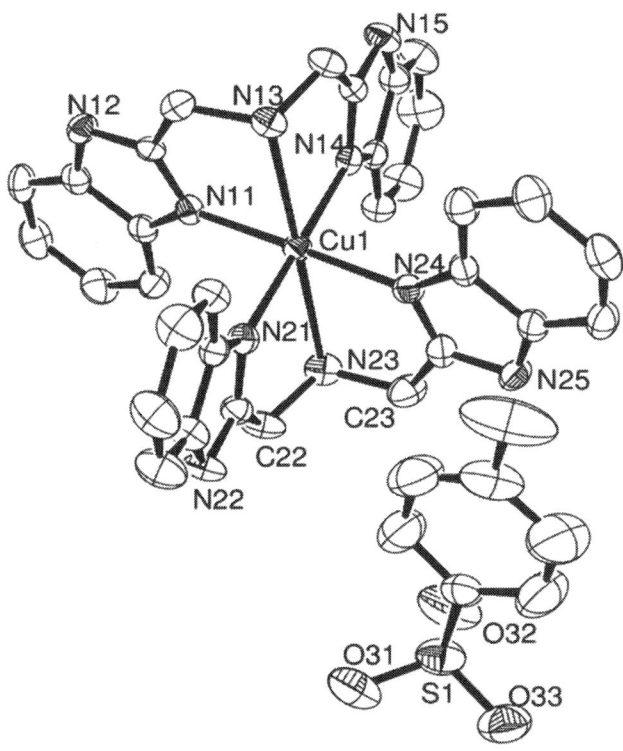

Figure 6: ORTEP diagram of [Cu (NNN)$_2$](RSO$_3$)$_2$·12H$_2$O showing the atom labeling scheme and ellipsoids drawn at 50% probability level. One toluene-4-sulfonate anion and twelve water molecules have been omitted for clarity.

Table 1: Selected crystallographic data for [Cu (NNN)$_2$](RSO$_3$)$_2$·12H$_2$O

[Cu(NNN)2](RSO3)2·12H2O. Compound	[Cu(NNN)2](RSO3)2·12H2O
Chemical formula	C54H86CuN10O18S2
Formulae weight	4514.66
Crystal color	Blue
Crystal system	Tetragonal
Space group	I4
Temperature (K)	200
Crystal size (mm3)	0.05 × 0.11 × 0.50

a (Å)	34.4165(11)
b (Å)	34.4165(11)
c (Å)	9.3238(4)
(°)	90
(°)	90
(°)	90
V (Å3)	11044.0(9)
Z	2
Dcalc (g/cm−3)	1.358
μ/mm-1	0.545
F(000)	4696
Wavelength (Å)	0.71073
Theta Min-Max (°)	1.9, 28.3
S	0.98
Tot., Uniq. Data, R(int)	91906, 13572, 0.095
Observed data [I > 2.0 sigma(I)]	8997
R	0.0507
Rw	0.1218

An X-ray crystal structure of the copper complex similar to $[Cu(NNN)_2](RSO_3)_2 \cdot 12H_2O$ was presented by Berends and Stephan [29] but no crystallographic parameters were reported because the complete refinement of the model was unsuccessful, and the ORTEP diagram presented was based on parameters obtained in "best" model. A similar nickel complex crystal structure has also been reported [30,31] but had a different arrangement of the donor groups, for example the aliphatic amine groups from each ligand were bonded trans to the benzimidazole group of another ligand. The similarity in the geometry of these base metal complexes afforded us to conclude that the lack of pH-metric separation with the use of bis((1-decylbenzimidazol-2-yl) methyl)amine as extractant (Section 3.1) is influenced by the lack of stereochemical "tailor making".

Table 2: Selected bond lengths (Å) and angles (°) for [Cu (NNN)$_2$] (RSO$_3$)$_2$·12H$_2$O

Bond lengths			
Cu1–N11	1.998(14)	Cu1-N21	2.045(14)
Cu1–N13	2.509(14)	Cu1-N23	2.547(14)
Cu1–N14	2.040(14)	Cu1-N24	2.012(15)
Bond angles			
N11-Cu1-N13	77.0(5)	N 1 3 - Cu1-N24	104.0(5)
N11-Cu1-N14	86.5(6)	N 1 4 - Cu1-N21	179.9(8)
N11-Cu1-N21	93.6(6)	N 1 4 - Cu1-N23	105.3(5)
N11-Cu1-N23	103.1(5)	N 1 4 - Cu1-N24	93.9(6)
N11-Cu1-N24	179.0(6)	N 2 1 - Cu1-N23	74.8(5)
N13-Cu1-N14	75.9(5)	N 2 1 - Cu1-N24	86.0(6)
N13-Cu1-N21	104.0(5)	N 2 3 - Cu1-N24	75.9(5)
N13-Cu1-N23	178.9(4)	C 2 2 - N23-C23	114.2(14)

CONCLUSIONS

The combination of two low pK$_a$ aromatic nitrogenous groups of benzimidazole with a strong aliphatic amine, in the design of a tridentate ligand, resulted in extraction curves that are pushed deep as a function of pH possibly due to the high complex formation constants of tridentate coordination and the relatively low protonation constants of the benzimidazole groups. This may compromise the stripping of the metal ions from the loaded organic phase through pH adjustment to lower pH. The Fe(III) rejection ability of this tridentate ligand in the range where the other later 3d metal ions extract is remarkable.

It can be concluded that the exploitation of the subtle stereochemical aspects of coordination for the extraction of base metals is lacking with tridentate ligands (at least those of the nature presented here). This leads us to propose the evaluation of a bidentate derivative, [2-methylaminomethyl-(1-R-benzimidazole), R = octyl or decyl], as a potential extractant of base metal ions. We have also demonstrated, for the first time, that the ion-association solvent extraction system studied here can be interpreted quantitatively for the complexation aspect of the formation of the cationic complex species to describe the two linear log D vs pH_e plots, and to extract information on coordination numbers from the extraction data using the respective slopes for each extraction curve.

ACKNOWLEDGEMENTS

The authors thank Mr F. Chindeka (DST/Mintek-NIC, Rhodes University Chemistry Department) for the microanalysis results. We also acknowledge Shell Chemicals (SA) (Pty) Ltd for supplying Shellsol 2325. We would also like to thank Mr A. S. Ogunlaja and Mr P. Kleyi for their assistance. For financial support, we thank the National Research Foundation (NRF-CPRR grant).

REFERENCES

1.   C. Kumar, S. K. Sahu and B. D. Pandey, "Prospects for Solvent Extraction Process in the Indian Context for Recovery of Base Metals. A Review," Hydrometallurgy, Vol. 103, No. 1-4, 2010, pp. 45-53. HUdoi:10.1016/j.hydromet.2010.02.016U.

2.   K. C. Sole, A. M. Feather and P. M. Cole, "Solvent Extraction in Southern Africa: An Update of Some Recent Hydrometallurgical Developments," Hydrometallurgy, Vol. 78, No. 1-2, 2005, pp. 52-78. HUdoi:10.1016/j.hydromet.2004.11.012U

3.   A. I. Okewole, N. P. Magwa and Z. R. Tshentu, "The Separation of Nickel(II) from Base Metal Ions Using 1-Octyl-2-(2'-pyridyl)imidazole as Extractant in a Highly Acidic Sulfate Medium," Hydrometallurgy, Vol. 121-124, 2012, pp. 81-89. HUdoi:10.1016/j.hydromet.2012.04.002U

4.   J. G. H. du Preez, J. Postma, S. Ravindran and B. J. A. M. van Brecht, "Nitrogen Reagents in Metal Ion Separation. Part VI. 2-(1'-Octylthiomethyl)pyridine as Extractant for Later 3d Transition Metal Ions," Solvent Extraction and Ion Exchange, Vol. 15, No. 1, 1997, pp. 79-96. HUdoi:10.1080/07366299708934 467U

5.   J. G. H. du Preez, "Recent Advances in Amines as Separating Agents for Metal Ions," Solvent Extraction and Ion Exchange, Vol. 18, No. 4, 2000, pp. 679-701. HUdoi:10.1080/073662900089 34703U

6.   G. D. Christian, "Analytical Chemistry," 6th Edition, John Wiley and Sons Inc, Hoboken, 2003, pp. 444-445.

7.   R. W. Hay, T. C. Clifford and P. Lightfoot, "Copper(II) and Zinc(II) Complexes of N,N-Bis(benzimidazole-2-ylmethyl)-amine. Synthesis, Formation Constants and the Crystal Structure of [ZnLCl]$_2$]·MeOH. Catalytic Activity of the Complexes in the Hydrolysis of the Phosphotriester 2,4-Dinitrophenyl Diethyl Phosphate," Polyhedron, Vol. 17, No. 20, 1998, pp. 3575-3581. HUdoi:10.1016/S0277-5387(98)00152-1U

8.   B. Kurzak, D. Kroczewska and J. Jezierska, "Ternary Copper(II) Complexes with Diethylenetriamine and -(or -) Alaninehydroxamic Acids in Water Solution," Polyhedron, Vol. 17, No. 11, 1998, pp. 1831-1841. HUdoi:10.1016/S0277-5387(97)00528-7U

9.   S. O. Bondareva, Y. I. Murinov and V. V. Lisitskii, "Extraction of Non-Ferrous Metals by Bisacylated Diethylenetriamine," Russian Journal of Inorganic Chemistry, Vol. 52, No. 5, 2007, pp. 796-799. HUdoi:10.1134/S0036023607050257U

10. J. G. H. du Preez, T. I. A. Gerber, W. Edge, V. L. V. Mtotywa and B. J. A. M. van Brecht, "Nitrogen Reagents in Metal Ion Separation. Part XI. The Synthesis and Extraction Behavior of a New Imidazole Derivative," Solvent Extraction and Ion Exchange, Vol. 19, No. 1, 2001, pp. 143-154. HUdoi:10.1081/SEI-100001379U

11. G. Wilkinson, J. A. McCleverty, and R. D. Gillard, Eds.,. Comprehensive Coordination Chemistry, Late Transition Elements, Vol. 5, Pergamon Press, 1987, pp. 596-681.

12. K. A. Allen, "Equilibrium between Didecylamine and Sulphuric Acid," Journal of Physical Chemistry, Vol. 60, No. 7, 1956, pp. 943-946. HUdoi:10.1021/j150541a027U

13. B. R. Reddy, S. V. Rao and K. H. Park, "Solvent Extraction Separation and Recovery of Cobalt and Nickel from Sulphate Medium using Mixtures of Tops 99 and TIBPS Extractants," Minerals Engineering, Vol. 22, No. 5, 2009, pp. 500-505. HUdoi:10.1016/j.mineng.2009.01.002U

14. J. V. Dagdigian and C. A. Reed, "A New Series of Imidazole-Thioether Chelating Ligands for Bioinorganic Copper," Inorganic Chemistry, Vol. 18, No. 9, 1979, pp. 2623-2626. HUdoi:10.1021/ic50199a058U

15. M. Haring, "A Novel Route to N-Substituted Heterocycles," Helvetica Chimica Acta, Vol. 42, 1957, pp. 1845- 1850.

16. Bruker SHELXTL Version 5.1. (Includes XS, XL, XP, XSHELL), Bruker AXS Inc., Madison, Wisconsin, USA, 1999.

17. A. L. Spek, "Single-Crystal Structure Validation with the Program PLATON," Journal of Applied Crystallography, Vol. 36, 2003, pp. 7-13. HUdoi:10.1107/S0021889802022112U

18. L. J. Farrugia, "ORTEP-3 for Windows: A Version of ORTEP-III with a Graphical User Interface (GUI)," Journal of Applied Crystallography, Vol. 30, 1997, p. 565. HUdoi:10.1107/S0021889897003117U

19. H. M. Irving and R. J. P. Williams, "The Stability of Transition Metal Complexes," Journal of the Chemical Society, 1953, pp. 3192-3210. HUdoi:10.1039/jr9530003192U

20. D. S. Flett, "Cobalt-Nickel Separation in Hydrometallurgy: A Review," Chemistry for Sustainable Development, Vol. 12, 2004, pp. 81-91.

21. L. Alderighi, P. Gans, A. Ienco, D. Peters, A. Sabatini and A. Vacca, "Hyperquad Simulation and Speciation (HySS): A Utility Program for the Investigation of Equilibria Involving Soluble and Partially Soluble Species," Coordination Chemistry Reviews, Vol. 184, No. 1, 1999, pp. 311-318. HUdoi:10.1016/S0010-8545(98)00260-4U

22. M. F. Perutz, "The Role of Aromatic Rings as HydrogenBond Acceptors in Molecular Recognition," Philosophical Transactions:

Physical Sciences and Engineering, Vol. 345, No. 1674, 1993, pp. 105-112. HUdoi:10.1098/rsta.1993.0122U

23. W. J. Geary, "The Use of Conductivity Measurements in Organic Solvents for the Characterisation of Coordination Compounds," Coordination Chemistry Reviews, Vol. 7, No. 1, 1971, pp. 81-122.

24. T. J. Lane, I. Nakagawa, J. L. Walker and A. J. Kandathil, "Infrared Investigation of Certain Imidazole Derivatives and Their Metal Chelates," Inorganic Chemistry, Vol. 1, No. 2, 1962, pp. 267-276. HUdoi:10.1021/ic50002a014U

25. J. Reedijk, "Pyrazoles and Imidazoles as Ligands. Part VI. Coordination Compounds of Metal(II) Perchlorates, Tetrafluoroborates and Nitrates Containing the Ligand Nn-Butylimidazole," Journal of Inorganic and Nuclear Chemistry, Vol. 33, No. 1, 1971, pp. 179-188. HUdoi:10.1016/0022-1902(71)80020-9U

26. E. S. Raper and J. L. Brooks, "Complexes of 1-Methylimidazoline-2-thione with Co(II) and Zn(II) Halides and Perchlorates," Journal of Inorganic and Nuclear Chemistry, Vol. 39, No. 12, 1977, pp. 2163-2166. HUdoi:10.1016/0022-1902(77)80387-4U

27. Z. Nakamoto, "Infrared and Raman Spectra of Inorganic and Coordination Compounds," 3rd Edition, John Wiley and Sons, New York, 1978, p. 239.

28. A. B. P. Lever, "Inorganic Electronic Spectroscopy," 2nd Edition, Elsevier, New York, 1984, pp. 554-557.

29. H. P. Berends and D. W. Stephan, "Copper(I) and Copper(II) Complexes of Biologically Relevant Tridentate Ligands," Inorganica Chimica Acta, Vol. 93, No. 4, 1984, pp. 173-178. HUdoi:10.1016/S0020-1693(00)88159-1U

30. J.-Y. Xu, W. Gu, L. Li, S.-P. Yan, P. Cheng, D.-Z. Liao and Z.-H. Jiang, "Synthesis and Crystal Structure of Nickel Complex of N,N-Bis(benzimidazol-2-yl-methyl)amine," Journal of Molecular Structure, Vol. 644, No. 1-3, 2003, pp. 23-27. HUdoi:10.1016/S0022-2860(02)00281-8U

31. P. Thangarasu, S. Bernès and C. Durán de Bazúa, "Bis [bis(benzimidazol-2-ylmethyl-N^3)amine-N]nickel(II) Dichloride", Acta Crystallographica Section C-Crystal Structure

Communications, Vol. 53, No. 11, 1997, pp. 1607- 1609. HUdoi:10.1107/S0108270197006513U

Chapter

8

Noble Metal Nanostructures Influence of Structure and Environment on Their Optical Properties

Ondřej Kvítek[1], Jakub Siegel[1], Vladimír Hnatowicz[2], and Václav Švorčík[1]

[1]Department of Solid State Engineering, Institute of Chemical Technology, 166 28 Prague, Czech Republic
[2]Nuclear Physics Institute, Academy of Sciences of the Czech Republic, 250 68 Rez, Czech Republic

ABSTRACT

Optical properties of nanostructured materials, isolated nanoparticles, and structures composed of both metals and semiconductors are broadly discussed. Fundamentals of the origin of surface plasmons as well as the surface plasmon resonance sensing are described and documented on a number of examples. Localized plasmon sensing and surface-enhanced Raman spectroscopy are subjected to special interest since those techniques are inherently associated with the direct application of plasmonic structures. The possibility of tailoring the optical properties of ultra-thin metal layers via controlling their shape and morphology by postdeposition annealing is documented. Special attention is paid to the contribution of bimetallic particles and layers as well as metal structures encapsulated in semiconductors and dielectrics to the optical response. The opportunity to tune the properties of materials over a large scale of values opens up entirely new application possibilities of optical active structures. The nature of surface plasmons predetermines noble metal nanostructures to be promising great materials for development of modern label-free sensing methods based on plasmon resonance—SPR and LSPR sensing.

INTRODUCTION

The crucial influence of the fine structure of materials on various mechanical, magnetic, and electronic properties was recognized at the end of 19th century. This line of thought seems to start with the pioneering work on the mechanical properties of iron alloys. These studies led to the conclusion that fine-scale microstructure retained after allotropic transformation of iron alloys gives martensite its hardness. The discovery of precipitation hardening in 1906 was the first observation suggesting that the correlation between microstructure and properties (originally proposed for ferrous alloys only) applies to nonferrous materials as well. Numerous observations in the subsequent years substantiated and generalized this view and led to the classification of the properties of solids with different types of chemical bonding into microstructure-sensitive and non-microstructure-sensitive ones. The physical understanding of the mechanisms by which ultrafine microstructures affect the properties of solids received a remarkable boost after the advent of the theory of lattice defects and the availability of new high-resolution characterization techniques such as electron and field ion microscopy. Both developments helped to elucidate the physical basis for understanding the correlation between the structure-sensitive properties and the microstructure of solids. Second period of developments in the area of nanomaterials started, when it was recognized that modification of the structure leads to generation of new atomic and electronic structures in the solids [1].

Bulk noble metals are known for their high thermal and electrical conductivity, specific mechanical properties, and high reflectivity of incident radiation. These properties are caused by their crystalline structure and presence of delocalized electrons—existence of electron gas. With decreasing thicknesses of metal films, new properties of the material can be observed. Optical properties are no exception, and a great shift can be seen in the appearance of a material of the same composition but different nanostructure. For example bulk gold is known as a shiny, yellow noble metal that does not tarnish. Smooth, thin films of the same metal appear opaque, and nanostructured surface covered with gold islands appears transparent with color changing from blue to red or even green depending on the structure size. Thus we can say that simply by changing the particle size or morphology of the material retaining its chemical composition, we can obtain structures with dramatically different properties. This property tuning by controlling the nanostructure of materials has proven to be very useful in specific applications.

PRINCIPLES

A good example of the tuning of the electronic structure of a material by modifying structural properties is CdS nanoparticles. By controlling the size of these semiconducting nanoparticles in the range of 2–5 nm, band gap of the material can be adjusted between 4.5 and 2.5 eV [4] The decreasing radius of the nanoparticles results in an increase of the band gap width with a decrease of the valence band maximum and an increase of the conductive band minimum, both contributing equally. This is very well displayed in the fluorescence color of the colloid solution (Figure 1(a)). It is normally considered that clusters of about 200 atoms have already the same unit cell and bond lengths as bulk material, but in the case of II–VI semiconductors, such as CdS, at least 10^4 atoms are needed for the bulk behavior to fully develop [5]. This is caused by a high ratio of the surface atoms in particles of this size. Increased influence of the irregular surface with high density of kinks, edges, and corners leads to presence of surface states which can trap electrons and holes and change optical and electronic properties of the material considerably. Shifts of band positions with varying particle size in optical absorption spectra of metallic nanoparticles resemble those of semiconductor clusters (compare Figures 1(a) and 1(b)). The origin of the resonant absorption in this case may, however, be quite different. Absorption due to electron interband transitions often roleplay (with decreasing size of a metal particle, band gap can be formed leading to transition of the properties of the material from metal to semiconductor or insulator), but the present delocalized conduction electrons which can be excited collectively into surface plasmons have also great influence on the mechanism of evolution and origin of certain absorption bands in the spectra [6].

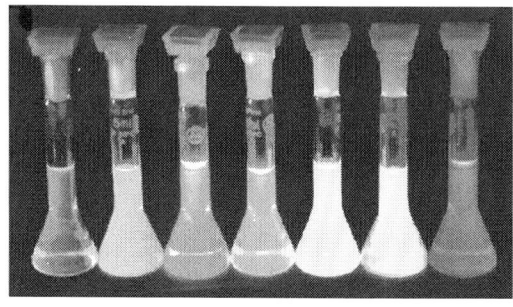

(a)

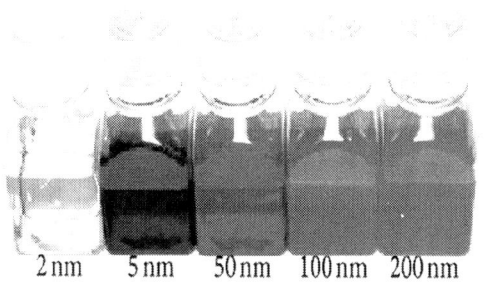

(b)

Figure 1: (a) Fluorescence of CdSe-CdS core-shell nanoparticles with a diameter of 1.7 nm (blue) up to 6 nm (red). (b) Size dependence of gold colloid optical absorption properties [2, 3].

Surface Plasmons and Surface Plasmon Resonance Sensing (SPR)

We have already stated that decreasing the size of the basic structural elements leads, among others, to changes of optical response of a material. A great deal of these changes is caused by increasing influence of certain electromagnetic surface modes—coherent fluctuations of electron charges on metal boundary called surface plasma oscillations or plasmons. Excitation of surface plasmons takes place, when the surface of the metal is exposed to incoming electrons or photons. Plasmons are strongly bound to the incident surface with their maximum intensity at the surface and disappear quickly with increasing distance from the surface. Therefore they are very sensitive to the surface properties. For photons to excite plasmons in the smooth metal surface, application of grating or attenuated total reflection (ATR) coupling is necessary to accomplish the need for appropriate wave vectors of the excited surface plasmon. On rough surfaces as well as gratings of appropriate parameters, surface plasmons can be reversely transformed into light. This light emission can play an important role in characterizing the roughness of the metal surface [7].

The propagation length of the surface plasmons in the metal surface as well as in surrounding medium is determined by their dissipation due to the internal and radiation damping. Internal damping is caused by excitation of electron-hole pairs at the Fermi level of the metal with following deexcitation producing phonons and thus heating the material (this process can be observed by photoacoustic measurements) and emission of the photoelectrons. Radiation damping consists in transformation of the evanescent wave of the surface plasmon into a plane wave. This process is characteristic of two-interface system such as the ATR device. The arrangement using ATR coupling to excite surface plasmons is nowadays well known and used in detection devices that evaluate the dependence of the incidence angle at which the surface plasmon resonance reaches its maximum (when frequency of coupled photons matches the frequency of the collective electron oscillations) on the medium surrounding the thin metal film (Figure 2). SPR detection methods prove to be very useful to study biological interactions in their natural state (or as close as possible), as there is, compared to standard methods of detection, no need for labeling.

Labels can affect interactions between the biomolecules, so label-free techniques are preferred. However, there are also problems with SPR sensors in comparison to other types of detection devices (mainly immunoassays), because of not-high-enough sensitivity of the SPR sensors and the necessity to integrate them into a complicated optic system to excite the surface plasmons, which hinders their application in common laboratory arrangements. There are a number of recent reviews particularly devoted to the field of SPR sensing and biosensing [8, 16, and 17].

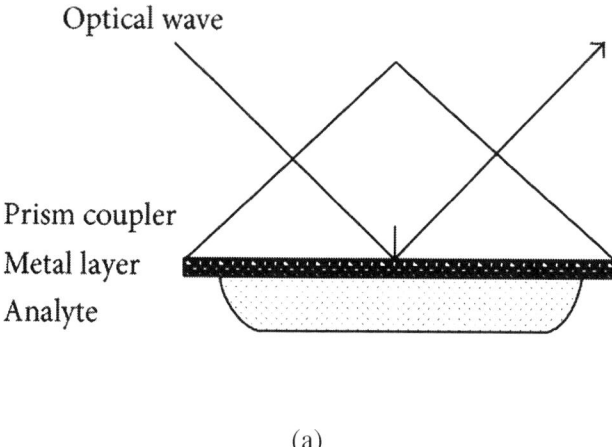

(a)

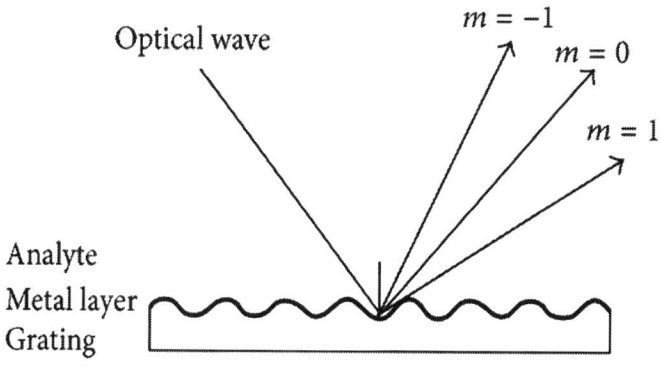

(b)

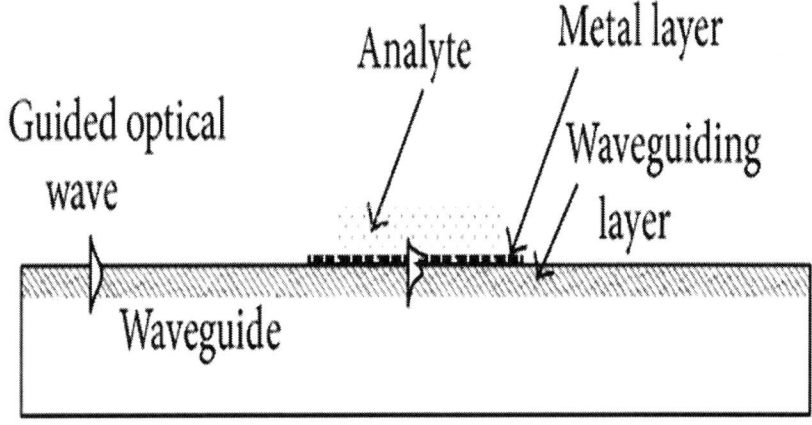

(c)

Figure 2: Most widely used configurations of SPR sensors: (a) prism coupler-based SPR system (ATR method); (b) grating coupler-based SPR system; (c) optical waveguide-based SPR system [8].

Localized Plasmons and Surface-Enhanced Raman Spectroscopy (SERS)

To eliminate the need for complicated coupling optics, nanostructured layers of isolated islands can be employed. In these structures, it is possible to excite surface plasmons by simple usage of incoming radiation with no need for ATR coupling or gratings, thanks to the naturally corrugated surface of the metal layer. The nature of these plasmons is a bit different from the plasmons observed on smooth surfaces as they are not propagating, but localized in the small area of confinement the nanostructure offers them. Rough metal surface was simulated using interlayers of CaF_2 of varying thicknesses, which provides surfaces of well-defined properties [18, 19]. Decreased phase velocity of surface plasmons calculated from this simulated interface is a consequence of multiple scatterings on the rough surface (scattering of the plasmon in new direction followed by rescattering into the original direction of the plasmon leading to reduction of the phase velocity at the preserved direction). Damping of the surface plasmons is determined by light emission into the air space. Experiments with monochromatic laser light at 1060 nm wavelength directed towards rough metal surfaces at 45° show that in the spectrum of scattered light of reflected beam, the signal for second harmonic of the incoming light (530 nm) is greatly amplified compared to the reflection on smooth metal surfaces [20]. This is caused by strong enhancement of the local electromagnetic field in the metal in resonance with the incident radiation. This effect has been used in development of the SERS method; it has been confirmed that the enhancement of the Raman signal is mainly due to an electromagnetic effect rather than chemical interaction between the adsorbed molecule and the metal. Since the discovery of this effect in 1970s, strongly enhanced Raman signals were verified for many different molecules attached to various rough metal surfaces (Figure 3). The estimated enhancement factors of the Raman signal range from modest factors of $10^3–10^5$ to enhancement factors of about $10^{10}–10^{11}$ for dye molecules in surface-enhanced resonance Raman spectroscopy (SERRS) experiments. Although single-molecule capabilities open up exciting perspectives for SERS as a tool in laboratory medicine and for basic research in biophysics, some of the experimental observations still create some controversy and are not yet completely understood. Together with the limitations attributed to the fact that the target molecules have to be attached to SERS-active substrates such as nanometer-sized silver or gold structures, it results in SERS not being a widely spread sensing method yet. In spite of these problems, this technique remains in focus of many research groups [9, 21–23].

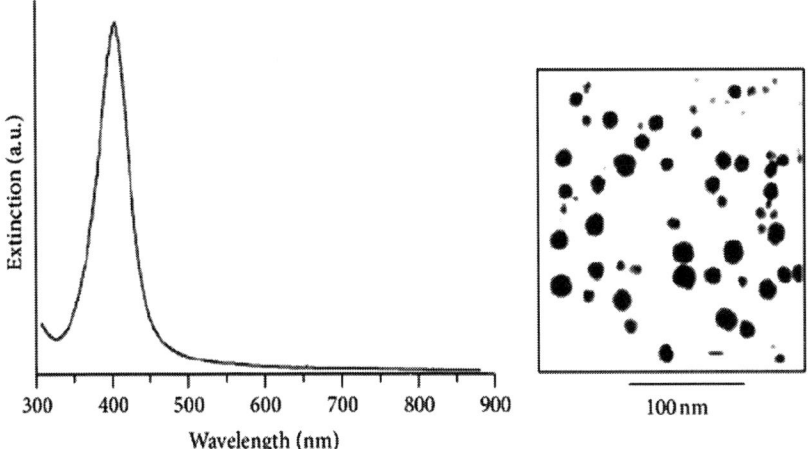

(a)

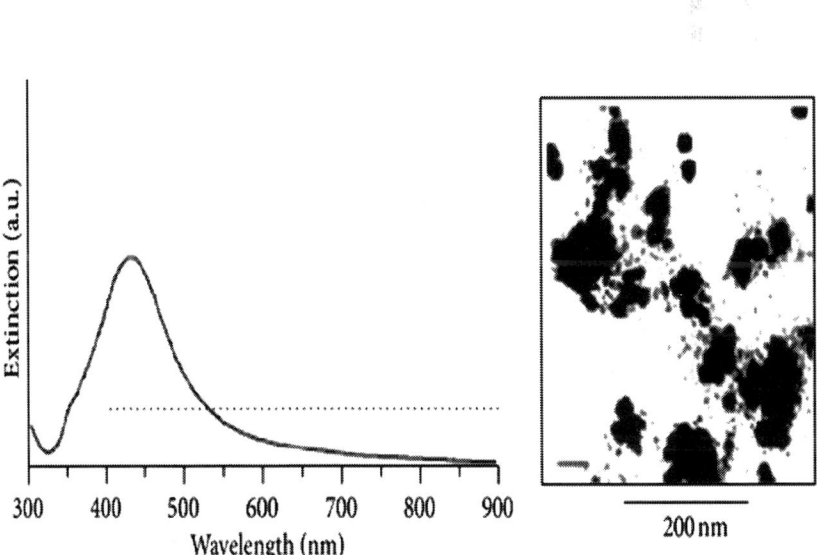

(b)

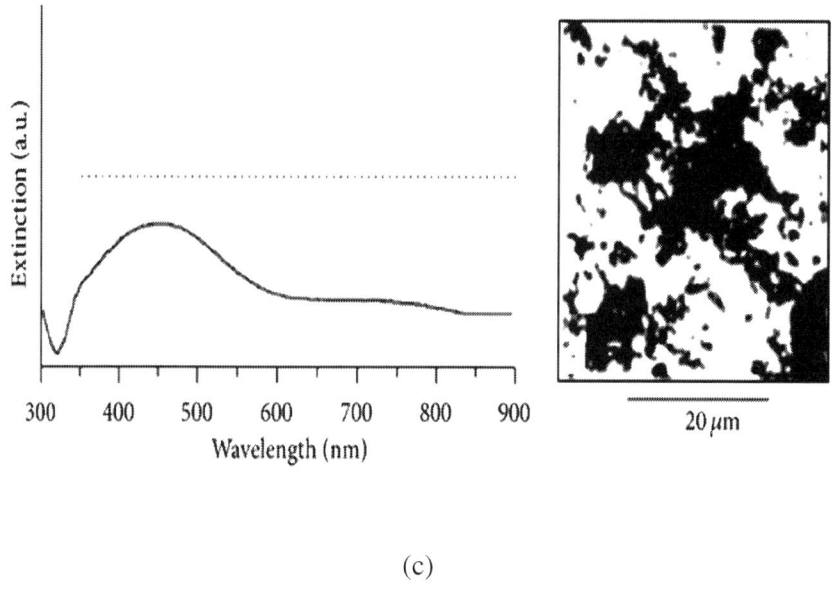

(c)

Figure 3: SERS-active colloidal silver particles in different aggregation stages, demonstrating the fractal nature of these structures together with the appropriate extinction curves [9].

Metal Nanoparticles and the Theory of Mie

The properties of localized surface plasmons have been studied for a long time, because the properties of nanostructured metal surfaces on insulating substrates and metal nanoparticles and colloid solutions are visibly influenced by the localized surface plasmon resonance effects. Already in 1908 Mie recognized that submicroscopical gold spheres differ in their optical properties from gold in the form of atoms and that it would be possible to study the optical absorption of colloid solutions to investigate how gold particles are composed. The theory Mie built upon the basis of Maxwell's equations is nowadays used to account for the frequency-dependent response of spherical metal particles to an electric field. According to this theory, the response is governed by real and imaginary parts of the dielectric function. Band positions, widths, and heights of plasmon oscillations, described in this theory as standing waves, depend sensitively on the dielectric function and on the particle radius [24]. Other factors influencing the surface plasmons in metal clusters are composition and shape [25]. Metal clusters with size above a critical value (usually about 5 nm radius) exhibit red-shift of the surface plasmon resonance. This shift increases with increasing size of the particle. Very small particles lose the metallic character, and their plasmons are suppressed. For example, the surface plasmon resonance wavelength of gold nanospheres can be tuned over 60 nm range by varying particle size between 10 and 100 nm [26]. This is extremely useful for optimization of surface enhanced-effects thermal treatments and also for maximization of the biosensing response of the nanostructures.

The red-shift of the surface plasmon resonance peak induced by a refractive index increase around metal nanostructures is the basis of the simplest sensing application of localized surface plasmons—the refractive index sensor. Noble metal nanoparticles and nanoobjects of various shapes immobilized on a smooth substrate surface are usually used for this purpose. Because localized surface plasmon resonance (LSPR) sensing is based on spectral peak shift, the precision that can be achieved with respect to changes in the refractive index depends on the sensitivity and the peak line width. Larger nanostructures tend to have high sensitivities, but their peaks are broadened by multipolar excitations and radiative damping. The LSPR shift is not strictly linear with the refractive index; therefore narrow spectral range of visible frequencies is usually probed in practice, where linearity of the LSPR shift with refractive index can be approximated. While LSPR sensors can detect changes in bulk refractive index, they are also capable of sensing localized to nanoscale distances from the nanostructure, because field enhancements due to LSPR decay rapidly with distance from the nanoparticle surface due to strong damping effects. This enables the observation of molecular interactions in the imminent vicinity of the nanoparticle surface [3, 27].

Evolution of the character of the material from molecular properties to metallic behaviour has been studied on visible range absorption spectra of clusters of sodium of different size. The atomic/molecular spectra observed at single-atom and 3-atom clusters change for 8-atom cluster in a smooth spectrum with a single absorption band, which shifts with further increasing size of the clusters to the longer wavelengths. Smooth thin films of sodium show monotonic spectra with no bands typical for the curved metal surfaces [28].

Besides the effect of the nanoparticle size, properties of the metal clusters are strongly influenced by their shape and structure as well. There has been a great deal of interest in shapes of nanoparticles that have sharp features or tips, which have been developed through both bottom-up and top-down methods. These include mainly silver and gold nanocubes, gold nanostars, silver nanotriangles, gold bipyramids and gold nanocrescents. One effect of the sharp tips is to produce a red-shift in the plasmon resonance, increasing the refractive index sensitivity [29–37].

THIN FILMS AND 2D STRUCTURES

A great basic example of dependence of optical properties of thin metal films on their thickness and structure is the case of Au nanolayers. Au is a good material for demonstration of the processes that take place during nanostructuring because of its great chemical stability. Švorčík et al. in [10] employed various analytical methods to determine the structure, thickness, morphology, and other properties of sputtered Au films. Annealing of the sputtered films at temperatures as low as 300°C leads to coalescence of the material and formation of nanoisland-like structure, which can be attributed to decrease of melting temperature of gold in the form of very thin film. The high surface energy between gold and glass substrate then leads to formation of discontinuous structure of particles spread quite evenly over the substrate surface, whose size depends on the thickness of the deposited Au layer prior to the annealing process.

The color of the continuous sputtered thin Au films ranges from blue to green with increasing thickness turning to the yellow color of the bulk gold for layers of thickness above 30 nm. The discontinuous layers of Au prepared by postdeposition annealing turn in color to the red shade ranging from light pink of the thinnest layers to dark violet for longer sputtering times (Figure 4). This effect is well described from the comparison of UV-Vis spectra and surface morphology of the samples. As annealing of the Au film leads to the formation of island-like structure (Figure 6), surface plasmons in the material become localized, and SPR band arises. While SPR band appears only for very short deposition times for unannealed samples, annealed samples of sputtered gold still show SPR band for layers of effective thickness of about 35 nm. The

SPR band of the annealed samples diminishes then for the samples with thickness above 40 nm, where the gold islands become too big to support localization and excitation of surface plasmons (Figure 5). Thus by this simple method based on the control of the Au film thickness prior to annealing, the nanostructured materials with wide variety of different optical absorption spectra in the visible region are produced.

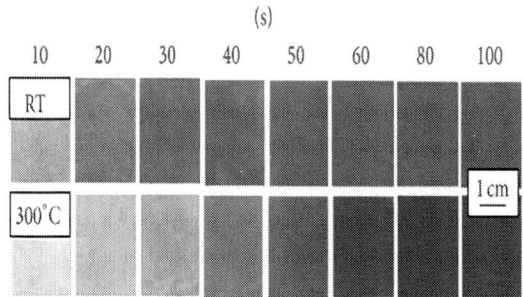

Figure 4: Photographs of the glass samples with gold structures sputtered for increasing times. The as-sputtered (RT) and annealed samples (300°C) are shown [10].

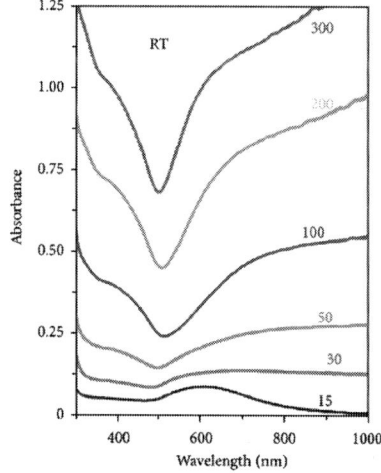

(a)

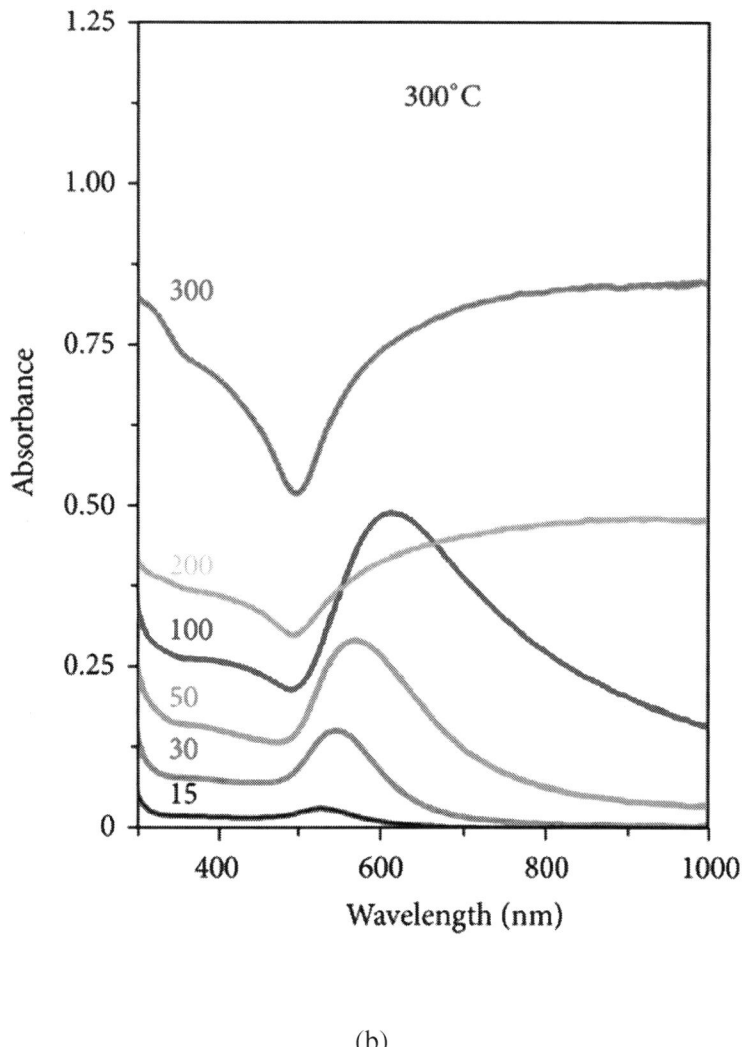

(b)

Figure 5: UV-Vis spectra of gold structures sputtered on glass before (RT) and after annealing (300°C) The numbers of lines refer to sputtering times in s [10].

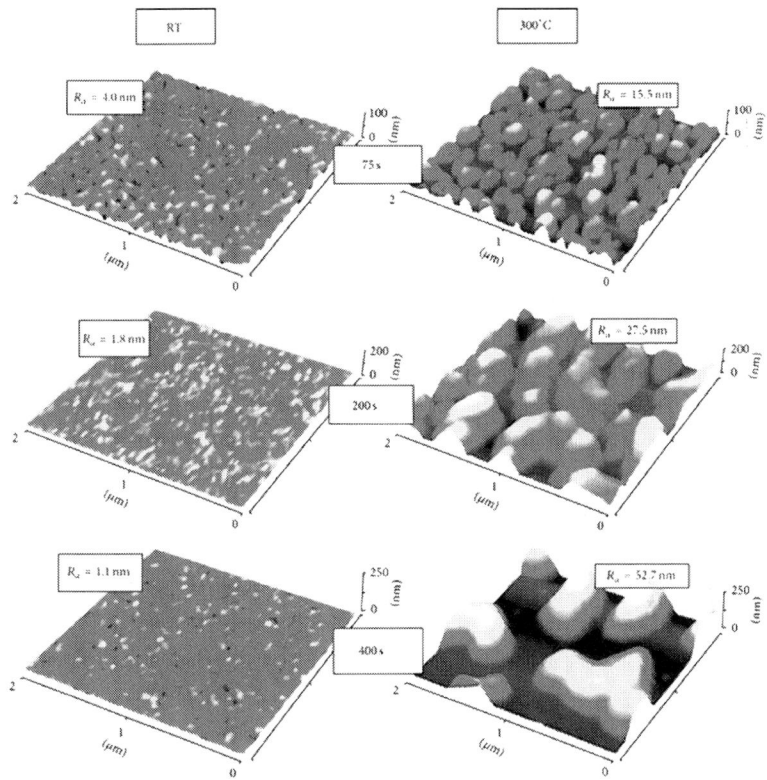

Figure 6: AFM images of gold structures sputtered for 75, 200, and 400 s on glass substrate before (RT) and after annealing (300°C) R$_a$ is the average surface roughness in nm [10].

The ellipsometric measurements of the sputtered thin Au films in [38] were used to determine refractive index and to calculate dielectric function. The real part of the dielectric function shows change from the shape typical for semiconductors and insulators to that typical for metals with increasing time of deposition. Annealing leads to occurrence of this transition at much longer deposition times. This measurement may raise some controversy about whether it is possible to calculate optical band gap from the absorption spectrum of the thin Au film. Tauc's method of optical band gap calculation has been used to calculate band gaps in semiconducting materials [39]. According to the ellipsometric measurements, the thin Au island-like film may possess semiconducting properties. The absorption spectrum of these systems is, however, strongly influenced by the SPR absorption band, which can make it difficult to apply this method of band gap calculation.

A different approach for the formation of nanostructured 2D assemblies of noble metals is to bind a pre-prepared colloid solution of nanoparticles to a beforehand treated substrate. Glass as the usual go-to choice for the substrate has a low affinity to noble metals. It is therefore necessary to modify its surface to make it possible to bind nanoparticles to it. For this purpose, salinization of the glass with various organic substances is often employed. In [11] glass plates were treated with aminopropyltrimethoxysilane and subsequently treated with colloid solutions of Au. UV-Vis spectra in this case showed strong dependence of the absorption on the time of the sample treatment in the colloid solution (Figure 7). Spectra with distinct SPR absorption bands were obtained, which indicates a successful binding of the nanoparticles. In [40] a similar method was used to compare the effectiveness of surface modification of the glass substrates by aminopropyltrimethoxysilane and mercaptopropyltrimethoxysilane. Viability of Au nanoparticles prepared by different methods (laser ablated, citrate and borohydride reduced) was also studied. Notably, it was found that aminopropyltrimethoxysilane is better for achieving salinization of the glass than mercaptopropyltrimethoxysilane, which is in contradiction to the usually accepted opinion that there is very strong interaction between noble metals and–SH groups. This effect can be, however, influenced by the type of used nanoparticle stabilization, because the nanoparticles prepared by laser ablation without any stabilizing agents were found to be unsuitable for the routine preparation of the samples. The structures prepared in this way were tested for their usability as a SERS substrate. Another example of employment of silanized glass substrates is in [12]. Subsequent modification of the immobilized nanoparticles by self-assembled monolayers of mercaptopropionic acid was used to bind biotin and this structure was used as a detection device for interrogation of interactions with fibrinogen. A significant change in the absorption spectrum maximum was observed after the introduction of fibrinogen (Figure 8). A different approach for employing self-assembled monolayers for Au nanoparticle immobilization was studied in [41]. Flat gold substrate surface was modified by 1, 10-decanedithiol, which formed structures with one–SH group bound to the surface and the other prepared to bind the subsequently introduced Au nanoparticle. A process of aggregation leading to structure of multiple layers of nanoparticles bound to the modified surface was observed. The obtained structures were tested for SPR detection of bovine serum albumin.

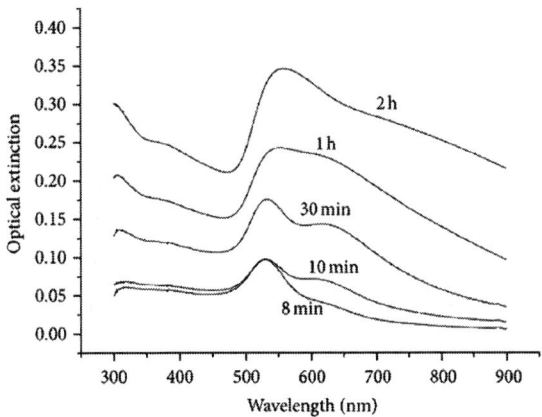

Figure 7: Effect of soaking time in the gold colloidal suspension on the UV absorbance of gold-decorated silanized glass substrates. Initial aminopropyl-trimethoxysilane concentration was 5% [11].

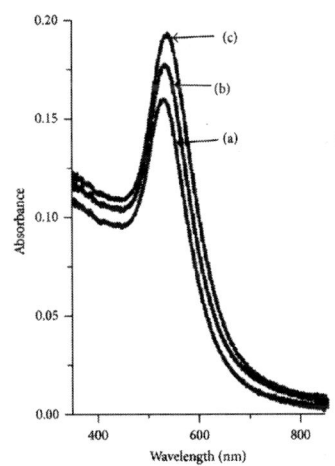

Figure 8: Absorbance spectrum of a mercaptopropionic acid functionalized colloidal gold monolayer on glass (a). Change in spectrum after 2 hours of incubation in (b) 10 and (c) 100 µg·mL solution of fibrinogen [12].

Examples of possible applications of such structured noble metal films are as was mentioned before in various analytical methods (particularly SERS), but also an attractive method to detect organic gases directly has been suggested in [42], where the dependence of SPR band position and shape on surrounding media refractive index has been employed using polymer overlayer able to absorb organic vapours. Shifts of SPR band have been also studied after direct absorption of organic gases on structured Au and Ag surfaces [43]. Another interesting example of possible application of thin Au films is in enhancing of organic solar cells, where both optical and electrical properties of the material are very important [44].

NANOPARTICLES

The ability to attach noble metal nanoparticles to surfaces may be very useful, especially if we realize that the prepared island-like 2D structure retains the properties of the original nanoparticle solution. If we were able to bind nanoparticles of various shapes, sizes, structure, and composition to solid surfaces without changing their character in the process, we could profit from advantages of both the specific particles and the 2D noble metal assembly. In this review, we will only mention several of the interesting recent observations concerning optical properties of noble metal nanoparticles, because there are exhaustive recent reviews focusing just on noble metal nanoparticles preparation and properties [26, 45, and 46].

Optical properties of metal nanoparticles are strongly dependent on their shape. Many recent developments have been made in regard to the control of the nanoparticle growth. In [47] properties and possible applications of various shapes (spheres, cubes, rods, and wires) of Au and Ag are discussed. Methods of preparation and shape control of gold nanoparticles with interesting comparison of gold nanoframes SPR absorption dependence on different solvents can be found in [13].

Measurement of optical absorption of colloid solutions may be used in the reverse sense as well—to determine concentration, size, and distribution of nanoparticles in the colloid solution. Correlation of width of the SPR peak of Au nanoparticles with their size and distribution was used to determine properties of colloid solutions in [48]. Interesting dependence of SPR absorption on measurement temperature is discussed as well. In a more recent study, mean free path correction was applied to classical Mie theory to characterize solutions of Au nanoparticles of size in the range of 5 nm to 100 nm. Equations as well as tables for particle size and concentration determination were supported [49].

An attractive method of large-scale preparation of monodisperse colloid solutions of very small Ag nanoparticles is discussed in [50]. In the synthesis of ultra-small Ag nanoparticles, oleylamine was used both as the reducing agent and surfactant, and oleic acid as the cosurfactant and cosolvent Ag nanoparticles as small as 1.7 nm were synthesized by controlling the growth by changing the heating rate UV-Vis spectra were measured for samples of solutions after different times of the synthesis to evaluate the state of the process (Figure 9). The measurements show an interesting progress in the nature of the spectra. During the last 2 min of the reaction time, a strong peak at 430 nm appears and grows rapidly, which is attributed to plasmonic scattering of Ag nanoparticles. This means that nucleation and growth took place in a very short period of time, less than 2 min at 180°C. In the controlled experiment, where oleic acid was not added to the reaction mixture, it was observed that the nanoparticles were formed at temperatures much lower, yielding only polydisperse silver particles. It seems that the coordination of oleic acid rather than oleylamine stabilizes the Ag monomers and enables burst of nucleation for the formation of uniform ultra-small nanoparticles.

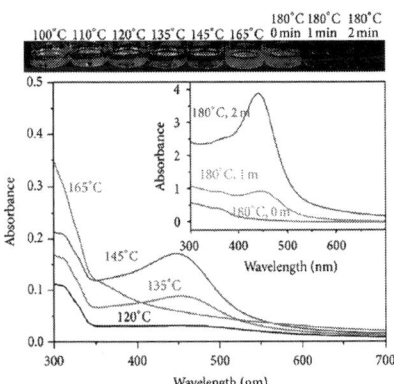

(a)

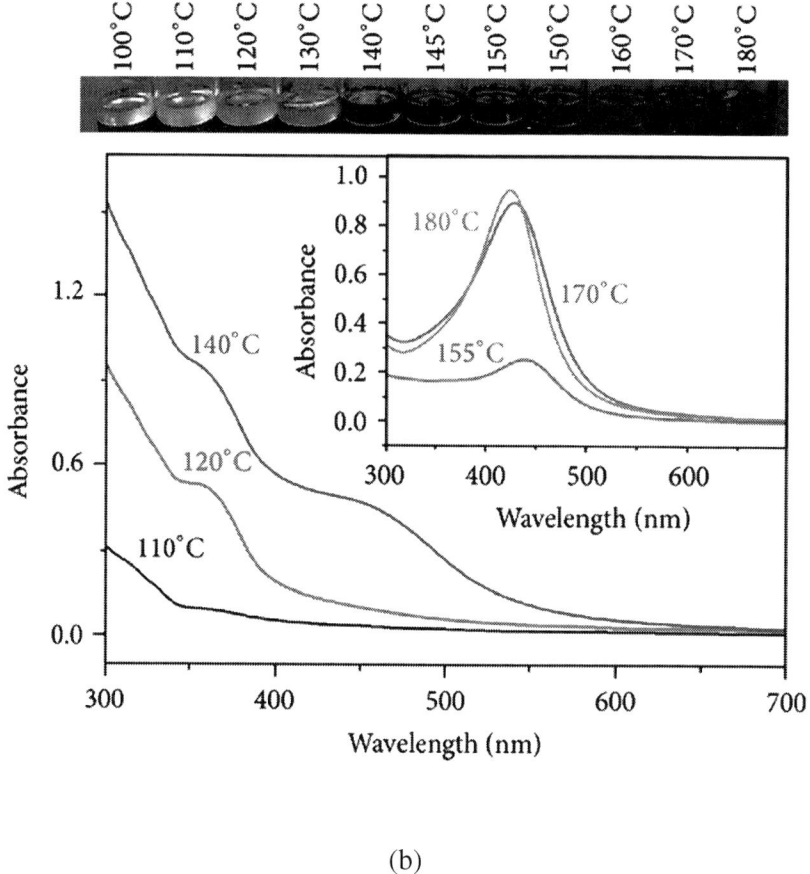

(b)

Figure 9: UV/Vis absorption spectra of Ag nanoparticles in various stages of preparation together with corresponding solution photographs. Reaction heating rate was (a) 10°C·min⁻¹ and (b) 1°C·min⁻¹ [13].

Another interesting method of Ag nanoparticles preparation uses chitosan as a reduction agent in a one-step preparation process to achieve a more environmentally friendly procedure [51]. Nanoparticles prepared by similar two-step process were studied in regard to their antibacterial activity [14]. UV-Vis extinction spectra in this case document the successful synthesis of silver nanoparticles (Figure 10). The spectra exhibit four characteristic peaks corresponding to different modes of plasmon excitation. These bands correspond to the formation of triangular Ag Nano plates.

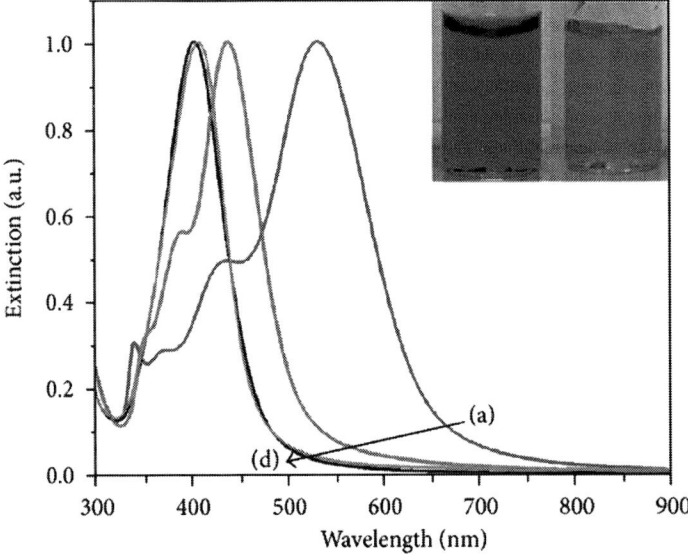

Figure 10: Normalized UV-Vis extinction spectra of silver nanoparticles enveloped in chitosan synthesized at 35°C with 16.5 mM trisodium citrate concentration (a) and with 4 mM trisodium citrate concentration (b) and at 0°C (c), respectively. A reference sample without chitosan at 0°C is shown in (d). The photographs correspond to silver colloids prepared at 35°C (left) and 0°C (right) [14].

Au nanoparticles are often used in biosensing applications. AuNPs are used as carriers for antibodies or other active molecules; methods using aggregation of AuNPs are based on shifts of the SPR absorption band with formation of AuNP clusters; nanoparticles are used to enhance various luminescence effects [52]. Recently, the specific ability of the particles to enhance electrochemiluminescence of peroxydisulfate used in constructing of biosensor has been reported [53].

BIMETALLIC PARTICLES AND LAYERS

We have already discussed how properties of materials change, when size of their structural elements decreases to nanoscopic scale without change of the composition of the material itself. Material properties are affected in a desirable way just by modification of basic structure features. This process is similar to that in which we create mixes of different compounds—by changing the ratio and distribution of the used materials we can tune the properties of the resulting structure. These two procedures can be combined, when we use two different metals to form a nanostructure or nanoparticle (Figure 11).

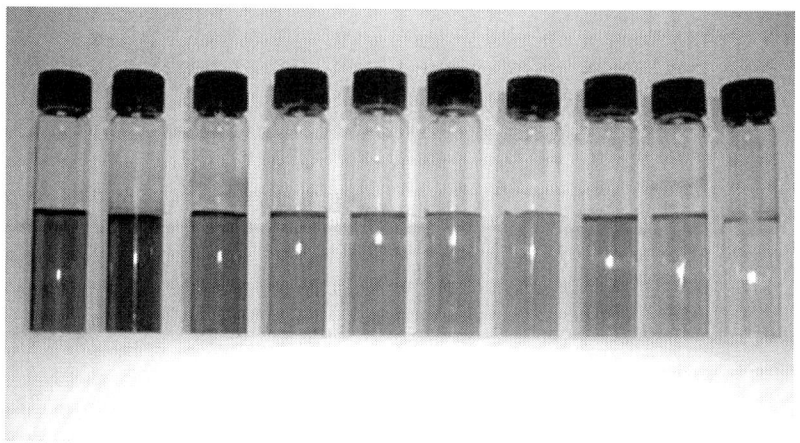

Figure 11: Colloidal solutions of Ag, Au and Ag-Au bimetallic nanoparticles [15].

Bimetallic core-shell Au/Ag nanoparticles can be prepared by two-step subsequent reduction of two metal salts in solution or by electroless deposition of silver onto gold nanoparticle cores. The formation of the bimetallic interface was found to increase sensitivity of LSPR sensors based on this structure [54]. Preparation of the core-shell nanoparticles in the reverse order than in the previous work (Ag/Au in this case) was accomplished by deposition of Au onto previously prepared Ag cores. Significant color change from yellow-amber to dark-amber to grey to grey-purple and finally to purple was observed during the Au deposition (Figure 12). Great attention was paid to the shift of the SPR peak position of the forming particles in regard to the amount of the Au introduced [55]. Absorption spectrum of alloy nanoparticles prepared by laser irradiation of mixture of colloidal solutions of Ag and Au nanoparticles (prepared beforehand by laser ablation in liquid environment) was found not to be a simple combination of the Au and Ag colloid solution spectra (Figure 13). Also, the dependence of the resulting spectra on the wavelength of the laser used for the preparation of the alloyed nanoparticles was observed [56]. A summary of preparation methods and means to achieve monodisperse colloid solutions of Au/Ag bimetallic nanoparticles was recently published [57].

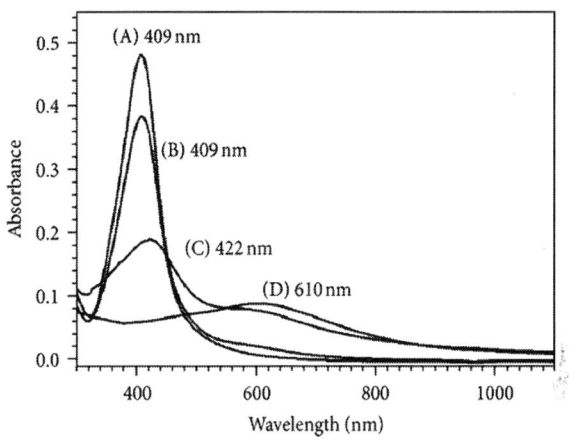

Figure 12: UV-Vis spectra for Ag and Ag/Au particles prepared with increasing Au content, as-synthesized Ag nanoparticles (A), 5% Au (B), 15% Au (C), 25% Au (D) [55].

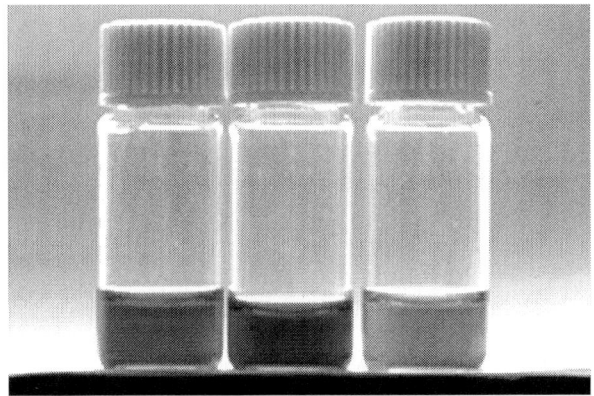

(a)

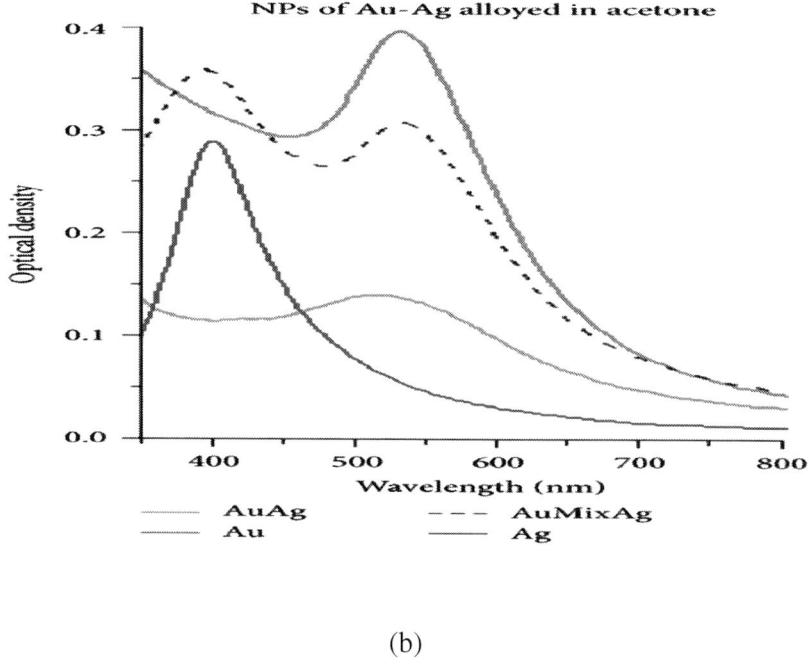

(b)

Figure 13: Photograph of mixture of colloidal solutions (a) of Au and Ag particles Initial mixture (left), after 2 hours (middle), and after 4 hours (right) of exposure to Cu laser radiation. (b) Spectra documenting formation of alloyed Au-Ag particles in acetone. Blue and red lines are the absorption spectra of initial NPs of Ag and Au, respectively. Dashed line is the absorption spectrum of the mixture of individual colloidal solutions prior to laser exposure. The green curve is the absorption spectrum of alloyed Au-AgNPs obtained by laser exposure of the mixture [56].

An example of specific application of optical properties of thin noble metal films is alloyed Pd hydrogen sensor. The intensity of light reflected from thin Pd film was found to be strongly dependent on the H_2 absorbed in the layer of the sensor in SPR arrangement [58]. The Pd/Ni alloy was found to be far more suited than pure palladium for the use in the sensor, owing to its greater mechanical strength and resistance to poisoning by other chemical species. Later, Au was introduced in the hydrogen sensing alloy, which can prevent certain phase transitions in the Pd layer and lead to better sensitivity of the resulting device [59].

METAL STRUCTURES ENCAPSULATED IN SEMICONDUCTORS AND DIELECTRICS

The dependence of specific optical absorption of nanostructured interfaces on the refractive index of surrounding medium brought to attention a possibility to use certain transparent solids to encapsulate the noble metal structures and improve their properties in this way. Therefore, nanocomposite materials consisted of noble metal thin layer structures encapsulated in semiconductors or dielectrics, or nanoparticles dispersed in such medium were developed. As with thin layers deposited on solid substrates, optical absorption of these composite materials can be modified by annealing. In this case, the annealing temperatures must be quite higher to achieve similar results, because the structure is stabilized by the surrounding solid [60].

The annealing temperature dependence of SPR peak in the UV-Vis spectra was studied in the case of Au structures encapsulated in TiO_2 (Figure 14). With increasing temperature of annealing, strong SPR peak of the nanostructured Au arises. The changes in the absorbance spectra can be detected already at 300°C, also confirming the first crystallization evidences through the Au clusters precipitation. Important to emphasis is that the SPR activity starts to develop for temperatures above 200°C and becomes higher at higher temperatures when the samples were annealed between 300 and 500°C, the optical density increases, together with a slight red-shift. The red-shift of the SPR peak can be attributed to both increasing size of the Au clusters and change in the value of the TiO_2 dielectric constant, as different ratios of amorphous TiO_2 to anatase are stable at different temperatures [61].

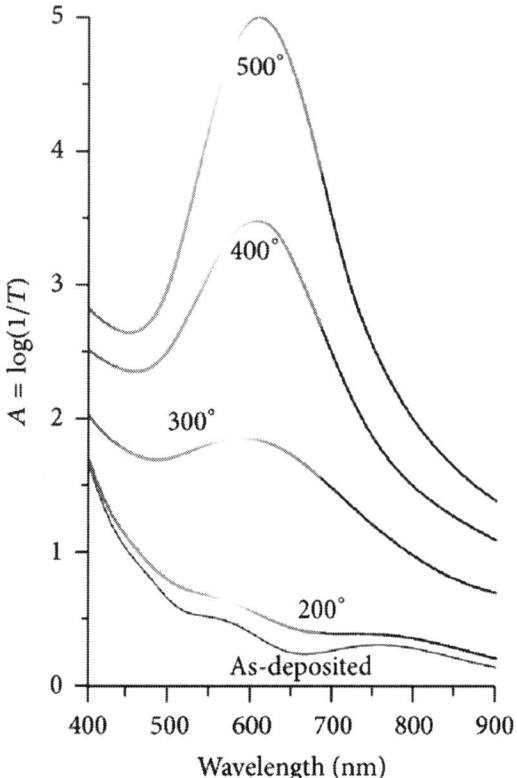

Figure 14: Absorbance spectra of Au:TiO$_2$ at different annealing temperatures [61].

Enhancement of NiO and ZnO thin film gas sensing by addition of Au nanoparticles into the structure was observed. Au nanoparticles embedded into the films by sol-gel method enabled the use of the structure as an optical sensor for pollutant gases—hydrogen, CO, and NO$_2$. The enhancement of the electrical detection was observed as well as improvement in dynamic characteristics of the sensor [62]. Another interesting structure for gas sensing was developed using ZnO:Ga nanowire. Adsorbing Au nanoparticles into the wire leads to increase of sensitivity of this sensor to CO by the order of magnitude [63].

Au nanoparticles were dispersed in CoO thin film using chemical solution approach combined with spin coating method. Interesting trend of evolution of SPR peak position was observed with increase of nanoparticle concentration (Figure 15). The peaks exhibit a red-shift and intensify with increasing Au content from 10 to 40 mol%, but show a blue shift and weaken from 40 to 60 mol% [64]. An interesting method of Au nanoparticles cultivation in borosilicate glasses of different viscosities was studied in [67]. The particle growth took place at 600°C. In more viscous glasses very small particles formed, with decreased viscosity, and larger particles showing SPR band appeared. Optical absorption spectra were used to calculate the size of formed nanoparticles and documented great influence of the glass viscosity on the kinetics of the Au nanoparticle formation process. Another intriguing type of matrix for nanoparticle embedding is polymers. Polytetrafluorethylene, poly (methyl methacrylate), and polyamide 6 were used in [68] as matrices for Ag Nano clusters. Nanocomposite films were prepared by vapour phase codeposition in high vacuum. The influence of various matrices and Ag clusters concentration on transmission spectrum of the material was tested.

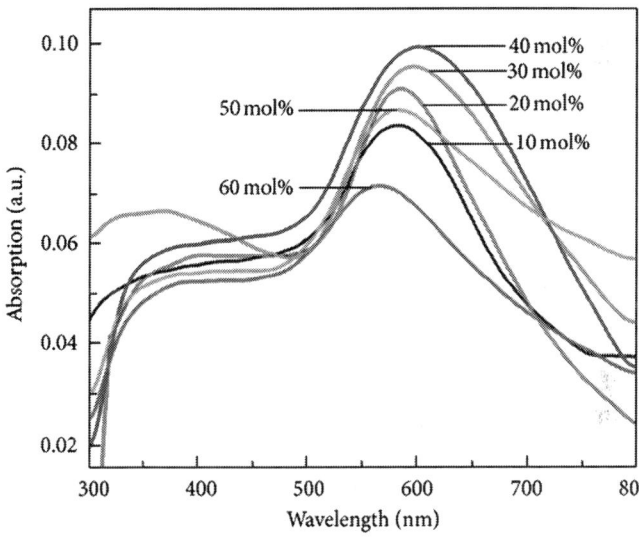

Figure 15: Optical absorption spectra of Au:CoO thin films with different contents of Au nanoparticles [64].

Au and Ag particles embedded in various semiconducting or dielectric matrices can be used in construction of nonlinear optic devices [69]. In regard to study of this phenomenon, Au/Ag alloyed particles were prepared in the SiO_2 matrix [70].

INFLUENCE OF SUBSTRATE ON THE SPR

Since plasmons propagate through all substances surrounding the metal nanostructures where they are excited, besides properties of over layers or surrounding atmosphere, the nature of the substrate on which a metal nanolayer is prepared has a crucial effect on the optical properties of the resulting material as well. The choice of a substrate can also have a great influence on the morphology of the growing nanostructure.

The properties of Au thin films deposited on glass and on SnO/In layer have been compared in [65]. The suppression of the SPR band of the nanostructured Au in the case of layers prepared on the said semiconducting layer can be seen in transmittance spectra (Figure 16). This effect can be explained by shifts in the stages of the thin film growth depending on the substrate qualities as well as influence of the conductivity and dielectric constant of the substrate on the plasmonic structure of the material.

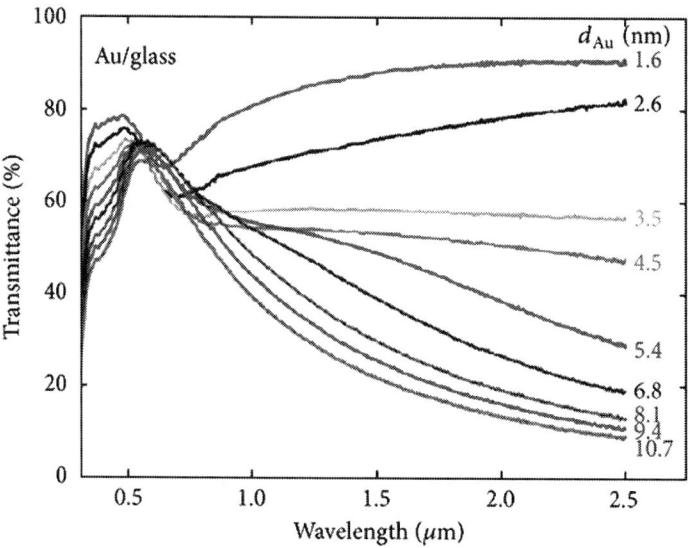

(a)

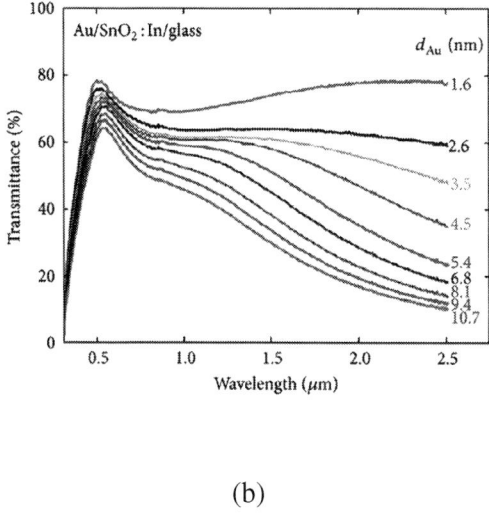

(b)

Figure 16: Spectral transmittance for Au films deposited on glass (a) and on SnO_2:In (b) [65].

The effect of the ratio of refractive indexes of overlayer and substrate was studied as well [66]. Nanostructured layers of Au islands were prepared on substrates of glass, quartz, and indium tin oxide (ITO) and subsequently coated with overlayers of silicon oxide, silicon nitride, ITO, diamond, or amorphous carbonated silicon. Postdeposition annealing was carried out to prepare island-like Au structure. The result was dependent on the quality of substrate with islands being the smallest (about 2 nm diameter) in the case of ITO and bigger (4 nm) in the case of glass and quartz (Figure 17).

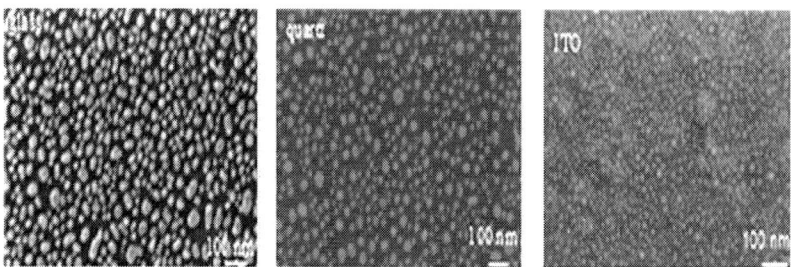

(a)

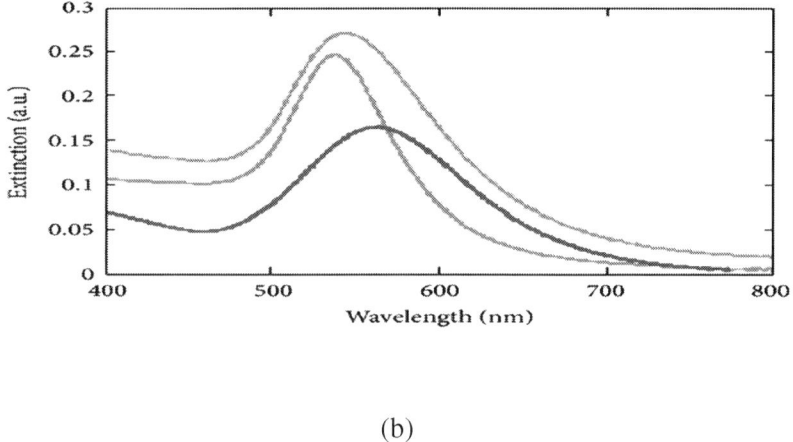

(b)

Figure 17: Scanning electron microscopy images (a) of glass/Au, quartz/Au, and ITO/Au nanostructured interfaces formed through thermal deposition of gold film and short high-temperature annealing. Corresponding UV/Vis extinction spectra are shown below (b). The lines in the spectra represent Au nanostructures on glass (green), quartz (red), and ITO (blue) [66].

An interesting type of substrate was proposed in [71], where pores of mesoporous SiO_2 were decorated by Au nanoparticles. An absorption peak at lower wavelength (around 470 nm), in addition to the Au SPR peak, induced by ambient aging and subsequent drying at 120°C was found. This can be caused by evolution of the structure of Au particles inside the pores.

Noble metals are well known to function as catalysts of redox reactions in gas sensors. The film of the noble metal has to be porous so that the detected gas could enter the sensitive layer. If we were able to control the structure of the noble metal catalyst film, we could increase the sensitivity of the detector greatly, especially in the low-temperature regime. One of the most studied oxidic semiconducting substrates that enable redox chemical reactions is recently TiO_2 as it enables various chemical reactions induced by visible light. The Au/Ag nanostructures prepared on its surface can significantly enhance its efficiency by tuning its band gap and mediating the electron transfers. Catalysts based on this principle can be used for degradation of polluting compounds [72], in synthesis of organic substances [73], or for generation of hydrogen for energetic purposes [74].

CONCLUSIONS

The classic assumption that the properties of a material are fully determined by its composition can nowadays be considered completely refuted with the rapid development of the nanotechnology. We can see that the properties of the material of the same composition can be altered in a wide range of values by simply controlling its nanostructure. With decreasing size of the structural elements of the material, quantum effects take place as the energy levels of electrons constituting the material are strongly influenced by the confinement due to the size of a nanoparticle. Confinement in such a way leads to strong intensification of collective oscillations of electrons, known as plasmons, having great effect on electrical and optical properties of the material. The propagating nature of those plasmons means that these properties can be affected by the media surrounding the nanostructure. This effect is a great promise in sensing applications. This also brings many more ways to tune the materials properties, as we can choose different substrates and over layers for the metal layers, nanostructures, and nanoparticles. Localized plasmons in island-like 2D structures and nanoparticles can be excited simply by incoming light and can produce great enhancement of electromagnetic fields, which is studied in regard to improvement of the Raman spectroscopy techniques.

The opportunity to tune properties of materials over a large scale of values opens up brand new possibilities in the applications of prepared structures. The nature of surface plasmons predetermines noble metal nanostructures to be great materials for development of modern label-free sensing methods based on plasmon resonance—SPR and LSPR sensing. The effect of electromagnetic field enhancement of localized plasmons is thoroughly studied by groups specializing in the development of SERS method. Increased reactiveness of noble metal nanoparticles has been found to be of great use in classic methods of gas sensing. Nanostructured noble metals even show great promises in improvement of organic solar cells effectiveness.

ACKNOWLEDGMENTS

This work was supported by the GACR under Project no. 108/12/G108. The authors thank Mrs. Oliva Kesselová for technical support.

REFERENCES

1. H. Gleiter, "Materials with ultrafine microstructures: retrospectives and perspectives," Nanostructured Materials, vol. 1, no. 1, pp. 1–19, 1992.

2. E. Roduner, "Size matters: why nanomaterials are different," Chemical Society Reviews, vol. 35, no. 7, pp. 583–592, 2006. · ·

3. K. M. Mayer and J. H. Hafner, "Localized surface plasmon resonance sensors," Chemical Reviews, vol. 111, no. 6, pp. 3828–3857, 2011. · ·

4. A. P. Alivisatos, "Semiconductor clusters, nanocrystals, and quantum dots," Science, vol. 271, no. 5251, pp. 933–937, 1996.

5. A. D. Yoffe, "Low-dimensional systems: quantum size effects and electronic properties of semiconductor microcrystallites (zero-dimensional systems) and some quasi-two-dimensional systems," Advances in Physics, vol. 42, no. 2, pp. 173–266, 1993.

6. E. Roduner, Nanoscopic Materials: Size Dependent Phenomena, RSC Publishing, Cambridge, UK, 2006.

7. H. Raether, Surface Plasmons on Smooth and Rough Surfaces and on Gratings, Springer, New York, NY, USA, 1988.

8. J. Homola, S. S. Yee, and G. Gauglitz, "Surface plasmon resonance sensors: review," Sensors and Actuators B, vol. 54, no. 1-2, pp. 3–15, 1999. · ·

9. K. Kneipp, H. Kneipp, I. Itzkan, R. R. Dasari, and M. S. Feld, "Surface-enhanced Raman scattering and biophysics," Journal of Physics Condensed Matter, vol. 14, no. 18, pp. R597–R624, 2002. · ·

10. V. Švorčík, O. Kvítek, O. Lyutakov, J. Siegel, and Z. Kolská, "Annealing of sputtered gold nano-structures," Applied Physics A, vol. 102, no. 3, pp. 747–751, 2011.

11. O. Seitz, M. M. Chehimi, E. Cabet-Deliry et al., "Preparation and characterisation of gold nanoparticle assemblies on silanised glass plates," Colloids and Surfaces A, vol. 218, no. 1–3, pp. 225–239, 2003. · ·

12. N. Nath and A. Chilkoti, "A colorimetric gold nanoparticle sensor to interrogate biomolecular interactions in real time on

a surface," Analytical Chemistry, vol. 74, no. 3, pp. 504–509, 2002. · ·

13.   J. Xiao and L. Qi, "Surfactant-assisted, shape-controlled synthesis of gold nanocrystals," Nanoscale, vol. 3, no. 4, pp. 1383–1396, 2011. · ·

14.   M. Potara, E. Jakab, A. Damert, O. Popescu, V. Canpean, and S. Astilean, "Synergistic antibacterial activity of chitosan-silver nanocomposites on Staphylococcus aureus," Nanotechnology, vol. 22, no. 13, Article ID 135101, 2011. · ·

15.   Q. Zhang, Y. N. Tan, J. Xie, and J. Y. Lee, "Colloidal synthesis of plasmonic metallic nanoparticles,"Plasmonics, vol. 4, no. 1, pp. 9–22, 2009. · ·

16.   D. R. Shankaran, K. V. Gobi, and N. Miura, "Recent advancements in surface plasmon resonance immunosensors for detection of small molecules of biomedical, food and environmental interest,"Sensors and Actuators B, vol. 121, no. 1, pp. 158–177, 2007. · ·

17.   E. Wijaya, C. Lenaerts, S. Maricot et al., "Surface plasmon resonance-based biosensors: from the development of different SPR structures to novel surface functionalization strategies," Current Opinion in Solid State and Materials Science, vol. 15, no. 5, pp. 208–224, 2011. · ·

18.   D. Hornauer, H. Kapitza, and H. Raether, "The dispersion relation of surface plasmons on rough surfaces," Journal of Physics D, vol. 7, no. 9, pp. L100–L102, 1974. · ·

19.   A. J. Braundmeier Jr. and E. T. Arakawa, "Effect of surface roughness on surface plasmon resonance absorption," Journal of Physics and Chemistry of Solids, vol. 35, no. 4, pp. 517–520, 1974.

20.   C. K. Chen, A. R. B. de Castro, and Y. R. Shen, "Surface-enchanced 2nd-harmonic generation," Physical Review Letters, vol. 46, no. 2, pp. 145–148, 1981.

21.   M. Moskovits, "Surface-enhanced spectroscopy," Reviews of Modern Physics, vol. 57, no. 3, pp. 783–826, 1985. · ·

22.   R. A. Tripp, R. A. Dluhy, and Y. Zhao, "Novel nanostructures for SERS biosensing," Nano Today, vol. 3, no. 3-4, pp. 31–37, 2008. · ·

23. B. Sharma, R. R. Frontiera, A.-I. Henry, E. Ringe, and R. P. Van Duyne, "SERS: materials, applications, and the future," Materials Today, vol. 15, no. 1-2, pp. 16–25, 2012. · ·

24. G. Mie, "Beitragezur Optik triiber Medien, speziell kolloidaler Metallösungen," Annals of Physics, vol. 25, no. 3, pp. 377–445, 1908.

25. L. M. Liz-Marzán, "Tailoring surface plasmons through the morphology and assembly of metal nanoparticles," Langmuir, vol. 22, no. 1, pp. 32–41, 2006. ·

26. M.-C. Daniel and D. Astruc, "Gold nanoparticles: assembly, supramolecular chemistry, quantum-size-related properties, and applications toward biology, catalysis, and nanotechnology," Chemical Reviews, vol. 104, no. 1, pp. 293–346, 2004. · ·

27. B. Sepúlveda, P. C. Angelomé, L. M. Lechuga, and L. M. Liz-Marzán, "LSPR-based nanobiosensors,"Nano Today, vol. 4, no. 3, pp. 244–251, 2009. · ·

28. U. Kreibig and M. Vollmer, Optical Properties of Metal Clusters, Materials Science, Springer, New York, NY, USA, 1995.

29. L. J. Sherry, S.-H. Chang, G. C. Schatz, R. P. Van Duyne, B. J. Wiley, and Y. Xia, "Localized surface plasmon resonance spectroscopy of single silver nanocubes," Nano Letters, vol. 5, no. 10, pp. 2034–2038, 2005. · ·

30. H. Chen, X. Kou, Z. Yang, W. Ni, and J. Wang, "Shape- and size-dependent refractive index sensitivity of gold nanoparticles," Langmuir, vol. 24, no. 10, pp. 5233–5237, 2008. · ·

31. C. L. Nehl, H. Liao, and J. H. Hafner, "Optical properties of star-shaped gold nanoparticles," Nano Letters, vol. 6, no. 4, pp. 683–688, 2006. · ·

32. L. J. Sherry, R. Jin, C. A. Mirkin, G. C. Schatz, and R. P. Van Duyne, "Localized surface plasmon resonance spectroscopy of single silver triangular nanoprisms," Nano Letters, vol. 6, no. 9, pp. 2060–2065, 2006. · ·

33. M. Liu and P. Guyot-Sionnest, "Mechanism of silver(I)-assisted growth of gold nanorods and bipyramids," Journal of Physical Chemistry B, vol. 109, no. 47, pp. 22192–22200, 2005. · ·

34. K. L. Kelly, E. Coronado, L. L. Zhao, and G. C. Schatz, "The optical properties of metal nanoparticles: the influence of size, shape,

and dielectric environment," Journal of Physical Chemistry B, vol. 107, no. 3, pp. 668–677, 2003. · ·

35. M. Liu, P. Guyot-Sionnest, T.-W. Lee, and S. K. Gray, "Optical properties of rodlike and bipyramidal gold nanoparticles from three-dimensional computations," Physical Review B, vol. 76, no. 23, Article ID 235428, 2007. · ·

36. J. M. McMahon, Y. Wang, L. J. Sherry et al., "Correlating the structure, optical spectra, and electrodynamics of single silver nanocubes," Journal of Physical Chemistry C, vol. 113, no. 7, pp. 2731–2735, 2009. · ·

37. R. Bukasov and J. S. Shumaker-Parry, "Highly tunable infrared extinction properties of gold nanocrescents," Nano Letters, vol. 7, no. 5, pp. 1113–1118, 2007. · ·

38. V. Švorčík, J. Siegel, P. Šutta et al., "Annealing of gold nanostructures sputtered on glass substrate,"Applied Physics a, vol. 102, no. 3, pp. 605–610, 2011.

39. J. Tauc, Amorphous and Liquid Semiconductors, Springer, New York, NY, USA, 1974.

40. N. Hajduková, M. Procházka, J. Štěpánek, and M. Špírková, "Chemically reduced and laser-ablated gold nanoparticles immobilized to silanized glass plates: preparation, characterization and SERS spectral testing," Colloids and Surfaces A, vol. 301, no. 1–3, pp. 264–270, 2007.

41. E. M. S. Azzam, A. Bashir, O. Shekhah et al., "Fabrication of a surface plasmon resonance biosensor based on gold nanoparticles chemisorbed onto a 1,10-decanedithiol self-assembled monolayer," Thin Solid Films, vol. 518, no. 1, pp. 387–391, 2009. · ·

42. T. Karakouz, A. Vaskevich, and I. Rubinstein, "Polymer-coated gold island films as localized plasmon transducers for gas sensing," Journal of Physical Chemistry B, vol. 112, no. 46, pp. 14530–14538, 2008. · ·

43. C.-S. Cheng, Y.-Q. Chen and C.-J. Lu, "Organic vapour sensing using localized surface plasmon resonance spectrum of metallic nanoparticles self-assemble monolayer," Talanta, vol. 73, no. 2, pp. 358–365, 2007. · ·

44. S. D. Yambem, A. Haldar, K.-S. Liao, E. P. Dillon, A. R. Barron, and S. A. Curran, "Optimization of organic solar cells with thin film Au as anode," Solar Energy Materials and Solar Cells, vol. 95, no. 8, pp. 2424–2430, 2011. · ·

45. T.-H. Tran and T.-D. Nguyen, "Controlled growth of uniform noble metal nanocrystals: aqueous-based synthesis and some applications in biomedicine," Colloids and Surfaces B, vol. 88, no. 1, pp. 1–22, 2011. · ·

46. E. C. Dreaden, A. M. Alkilany, X. Huang, C. J. Murphy, and M. A. El-Sayed, "The golden age: gold nanoparticles for biomedicine," Chemical Society Reviews, vol. 41, no. 7, pp. 2740–2779, 2012. · ·

47. A. Seyed-Razavi, I. K. Snook, and A. S. Barnard, "Origin of nanomorphology: does a complete theory of nanoparticle evolution exist?" Journal of Materials Chemistry, vol. 20, no. 3, pp. 416–421, 2010. · ·

48. S. Link and M. A. El-Sayed, "Size and temperature dependence of the plasmon absorption of colloidal gold nanoparticles," Journal of Physical Chemistry B, vol. 103, no. 21, pp. 4212–4217, 1999.

49. W. Haiss, N. T. K. Thanh, J. Aveyard, and D. G. Fernig, "Determination of size and concentration of gold nanoparticles from UV-Vis spectra," Analytical Chemistry, vol. 79, no. 11, pp. 4215–4221, 2007. · ·

50. J. Park, S. G. Kwon, S. W. Jun, B. H. Kim, and T. Hyeon, "Large-scale synthesis of ultra-small-sized silver nanoparticles," Journal of Physical Chemistry A, vol. 13, no. 10, pp. 2540–2543, 2012.

51. A. Murugadoss and A. Chattopadhyay, "A "green" chitosan-silver nanoparticle composite as a heterogeneous as well as micro-heterogeneous catalyst," Nanotechnology, vol. 19, no. 1, Article ID 015603, 2008. · ·

52. X. Cao, Y. Ye, and S. Liu, "Gold nanoparticle-based signal amplification for biosensing," Analytical biochemistry, vol. 417, no. 1, pp. 1–16, 2011. · ·

53. B. H. Gao, S. N. Ding, O. Kargbo, Y. H. Wang, Y. M. Sun, and S. Cosnier, "Enhanced electrochemiluminescence of peroxydisulfate by electrodeposited Au nanoparticles and its biosensing application via integrating biocatalytic precipitation

using self-assembly bi-enzymes," Journal of Electroanalytical Chemistry, vol. 703, pp. 9–13, 2013.

54. A. Csáki, S. Berg, N. Jahr et al., "Plasmonic nanoparticles— noble material for sensoric applications," inGold Nanoparticles: Properties, Characterization and Fabrication, NOVA Science, Huntington, NY, USA, 2009.

55. D. Mott, N. T. B. Thuy, Y. Aoki, and S. Maenosono, "Synthesis of size and shape controlled silver nanoparticles coated by a thin layer of gold and their use as ultrasensitive biomolecular probes," inProceedings of the Materials Research Society in Functional Materials and Nanostructures for Chemical and Biochemical Sensing, vol. 1253, pp. 652–663, April 2010.

56. G. A. Shafeev, "Laser synthesis of gold nanoparticles and the control over their properties," in Gold Nanoparticles: Properties, Characterization and Fabrication, NOVA Science, New York, NY, USA, 2009.

57. Q. Zhang, J. Xie, Y. Yu, and J. Y. Lee, "Monodispersity control in the synthesis of monometallic and bimetallic quasi-spherical gold and silver nanoparticles," Nanoscale, vol. 2, no. 10, pp. 1962–1975, 2010. · ·

58. B. Chadwick, J. Tann, M. Brungs, and M. Gal, "A hydrogen sensor based on the optical generation of surface plasmons in a palladium alloy," Sensors and Actuators B, vol. 17, no. 3, pp. 215–220, 1994.

59. S. Nazarpour, A. Cirera, and M. Varela, "Material properties of Au-Pd thin alloy films," Thin Solid Films, vol. 518, no. 20, pp. 5715–5719, 2010. · ·

60. L. Armelao, D. Barreca, G. Bottaro et al., "Recent trends on nanocomposites based on Cu, Ag and Au clusters: a closer look," Coordination Chemistry Reviews, vol. 250, no. 11-12, pp. 1294–1314, 2006. · ·

61. M. Torrell, L. Cunha, A. Cavaleiro, E. Alves, N. P. Barradas, and F. Vaz, "Functional and optical properties of Au:TiO_2 nanocomposite films: the influence of thermal annealing," Applied Surface Science, vol. 256, no. 22, pp. 6536–6542, 2010. · ·

62. E. Della Gaspera, M. Guglielmi, A. Martucci, L. Giancaterini, and C. Cantalini, "Enhanced optical and electrical gas sensing

response of sol-gel based NiO-Au and ZnO-Au nanostructured thin films," Sensors and Actuators B, vol. 164, no. 1, pp. 54–63, 2012. · ·

63.   S.-J. Chang, T.-J. Hsueh, I.-C. Chen and B.-R. Huang, "Highly sensitive ZnO nanowire CO sensors with the adsorption of Au nanoparticles," Nanotechnology, vol. 19, no. 17, Article ID 175502, 2008. · ·

64.   S.-J. Wang, B.-P. Zhang, L.-P. Yan, and W. Deng, "Microstructure and optical absorption properties of Au-dispersed CoO thin films," Journal of Alloys and Compounds, vol. 509, no. 19, pp. 5731–5735, 2011. · ·

65.   P. C. Lansåker, K. Gunnarsson, A. Roos, G. A. Niklasson, and C. G. Granqvist, "Au thin films deposited on SnO_2: in and glass: substrate effects on the optical and electrical properties," Thin Solid Films, vol. 519, no. 6, pp. 1930–1933, 2011. · ·

66.   S. Szunerits and R. Boukherroub, "Short and long range sensing on plasmonic nanostructures coated with oxide-based dielectrics," in Gold Nanoparticles: Properties, Characterization and Fabrication, NOVA Science, New York, NY, USA, 2009.

67.   J. Vosburgh and R. H. Doremus, "Optical absorption spectra of gold nano-clusters in potassium borosilicate glass," Journal of Non-Crystalline Solids, vol. 349, no. 1–3, pp. 309–314, 2004. · ·

68.   H. Takele, H. Greve, C. Pochstein, V. Zaporojtchenko, and F. Faupel, "Plasmonic properties of Ag nanoclusters in various polymer matrices," Nanotechnology, vol. 17, no. 14, pp. 3499–3505, 2006. · ·

69.   J. P. Huang and K. W. Yu, "Enhanced nonlinear optical responses of materials: composite effects,"Physics Reports, vol. 431, no. 3, pp. 87–172, 2006. · ·

70.   P. Sangpour, O. Akhavan, and A. Z. Moshfegh, "The effect of Au/Ag ratios on surface composition and optical properties of co-sputtered alloy nanoparticles in Au-Ag:SiO_2 thin films," Journal of Alloys and Compounds, vol. 486, no. 1-2, pp. 22–28, 2009. · ·

71.   G. Fu, W. Cai, Y. Gan, and J. Jia, "An ambience-induced optical absorption peak for Au/SiO_2mesoporous assembly," Chemical Physics Letters, vol. 385, no. 1-2, pp. 15–19, 2004. · ·

72.　D. Gong, W. C. J. Ho, Y. Tang et al., "Silver decorated titanate/ titania nanostructures for efficient solar driven photocatalysis," Journal of Solid State Chemistry, vol. 189, pp. 117–122, 2012. · ·

73.　R. Kaur and B. Pal, "Size and shape dependent attachments of Au nanostructures to TiO_2 for optimum reactivity of Au-TiO_2 photocatalysis," Journal of Molecular Catalysis A, vol. 355, pp. 39–43, 2012. · ·

74.　H. Yuzawa, T. Yoshida, and H. Yoshida, "Gold nanoparticles on titanium oxide effective for photocatalytic hydrogen formation under visible light," Applied Catalysis B, vol. 115-116, pp. 294–302, 2012. · ·

Citations

CHAPTER 1

Evarice Yama Nzoma, Alain Guillet, and Philippe Pareige, "Nano-structured Multifilamentary Carbon-Copper Composites: Fabrication, Microstructural Characterization, and Properties," Journal of Nanomaterials, vol. 2012, Article ID 360818, 11 pages, 2012. doi:10.1155/2012/360818.

CHAPTER 2

A. Dhanapal, S. Rajendra Boopathy, V. Balasubramanian, K. Chidambaram, and A. R. Thoheer Zaman, "Experimental Investigation of the Corrosion Behavior of Friction Stir Welded AZ61A Magnesium Alloy Welds under Salt Spray Corrosion Test and Galvanic Corrosion Test Using Response Surface Methodology," International Journal of Metals, vol. 2013, Article ID 317143, 17 pages, 2013. doi:10.1155/2013/317143.

CHAPTER 3

N. Kumar, A. Jyothirmayi, K. R. C. Soma Raju, V. Uma, and R. Subasri, "One-Step Anodization/Sol-Gel Deposition of Ce^{3+} -Doped Silica-Zirconia Self-Healing Coating on Aluminum," ISRN Corrosion, vol. 2013, Article ID 424805, 8 pages, 2013. doi:10.1155/2013/424805.

CHAPTER 4

M. Abdus Salam, Bawadi Abdullah, and Suriati Sufian, "Hydrogenated Microstructure and Its Hydrogenation Properties: A Density Functional Theory Study," Journal of Nanomaterials, vol. 2014, Article ID 749804, 7 pages, 2014. doi:10.1155/2014/749804.

CHAPTER 5

Bukhanovsky, V., Rudnytsky, M. and Mamuzich, I. (2014) Microlayered Composite Materials on Basis of Copper, Refractory, Rare-Earth Metals, and Carbon for Electrical Contacts and Electrodes International Journal of Nonferrous Metallurgy, 3, 18-27. doi: 10.4236/ijnm.2014.32003.

CHAPTER 6

A. I. Alateyah, H. N. Dhakal, Z. Y. Zhang, and B. Aldousiri, "Low Velocity Impact and Creep-Strain Behaviour of Vinyl Ester Matrix Nanocomposites Based on Layered Silicate," International Journal of Polymer Science, vol. 2014, Article ID 541096, 10 pages, 2014. doi:10.1155/2014/541096.

CHAPTER 7

Magwa, N. , Hosten, E. , Watkins, G. and Tshentu, Z. (2012) An Explor-atory Study of Tridentate Amine Extractants: Solvent Extraction and Co-ordination Chemistry of Base Metals with *Bis*((1*R*-benzimidazol-2-yl) methyl)amine. *International Journal of Nonferrous Metallurgy*, 1, 49-58. doi: 10.4236/ijnm.2012.13007.

CHAPTER 8

Ondřej Kvítek, Jakub Siegel, Vladimír Hnatowicz, and Václav Švorčík, "Noble Metal Nanostructures Influence of Structure and Environment on Their Optical Properties," Journal of Nanomaterials, vol. 2013, Ar-ticle ID 743684, 15 pages, 2013. doi:10.1155/2013/743684.

Index